普通高等学校"十四五"规划机械类专业精品教材

工程图学习题集（中英对照）

主 编 庄 宏 叶福民

华中科技大学出版社
中国·武汉

内 容 简 介

本书是作者根据多年的教学经验并结合近年来教学改革实践而编写的。全书共分 9 章,主要内容包括制图基本知识、几何元素的投影、立体及其交线、组合体、图样画法、标准件与常用件、零件图、装配图、计算机绘图等。另外,本书还配有大量数字资源,包括参考答案、立体模型及演示动画,读者可通过微信端扫描书中二维码获取。

本书可作为高等学校机械类、近机械类及其他工程类各专业开设的"工程图学"课程的教材,也可作为工程设计人员的参考用书。

图书在版编目(CIP)数据

工程图学习题集:中英对照/庄宏,叶福民主编. —武汉:华中科技大学出版社,2023.6(2024.7重印)
ISBN 978-7-5680-9476-4

Ⅰ.①工… Ⅱ.①庄… ②叶… Ⅲ.①工程制图-高等学校-习题集-汉、英 Ⅳ.①TB23-44

中国国家版本馆 CIP 数据核字(2023)第 090318 号

工程图学习题集(中英对照)
Gongchengtuxue Xitiji(Zhong-Ying Duizhao)

庄 宏 叶福民 主编

策划编辑:王 勇
责任编辑:姚同梅
封面设计:原色设计
责任监印:周治超

出版发行:华中科技大学出版社(中国·武汉)　　电话:(027)81321913
　　　　　武汉市东湖新技术开发区华工科技园　　邮编:430223
录　　排:武汉三月禾文化传播有限公司
印　　刷:武汉市洪林印务有限公司
开　　本:889mm×1194mm　1/16
印　　张:13.5
字　　数:410千字
版　　次:2024年7月第1版第2次印刷
定　　价:39.80元

本书若有印装质量问题,请向出版社营销中心调换
全国免费服务热线:400-6679-118　竭诚为您服务
版权所有　侵权必究

Preface

This book is written in accordance with the requirements of engineering graphics courses in colleges and universities, based on the authors' years of teaching experience and the practice of teaching reform in recent years. The book is oriented by practical application of engineering, focuses on training students, awareness of standardization and ability to draw and read engineering drawings. And the book adopts the latest national standards.

The book is divided into 9 chapters, the number of exercises in each chapter and the difficulty are appropriate. The main contents include basic knowledge of drawing, projection of geometric elements, solids and their intersections, composite solids, general principles of representation, standard parts and commonly used parts, detail drawing, assembly drawings, computer drawing, etc. The entire book proceeds from the shallow to the deep, step by step, and it is highly targed and emphasizes practicality and universality. This book is also equipped with digital resources, including detailed problem solutions, reference answers, corresponding three-dimensional models and animation demonstrations, which are beneficial for students to systematically grasp knowledge and improve their abilities.

This book can be used as a teaching material for the engineering graphics course of mechanical, near mechanical and other engineering majors in colleges and universities, and also as a reference book for engineering designers.

The book is edited by Zhuang Hong and Ye Fumin. Chapters 5,6 and 9 are written by Ye Fumin and the remaining chapters by Zhuang Hong.

I would like to express my sincere gratitude to the First Class Undergraduate Professional Construction Fund of Jiangsu University of Science and Technology for supporting the writing of this book. Due to the limited level of editors, there are inevitably shortcomings in the book. Suggestions for improvement will be gratefully received.

Zhuang Hong

2023. 2

前 言

本书是作者按照高等学校工程图学课程的要求，根据多年的教学经验并结合近年来教学改革实践而编写的。本书以实际工程应用为导向，着重培养学生的标准化意识及绘制和阅读工程图样的能力。全书采用了最新的国家标准。

本书共分9章，每章题量适当，难度适宜。主要内容包括制图基本知识、几何元素的投影、立体及其交线、组合体、图样画法、标准件与常用件、零件图、装配图、计算机绘图等。全书内容由浅入深，循序渐进，针对性强，注重实用性和通用性。本书配有二维码资源，包括详细题解、参考答案、相应的立体模型及动画演示，有利于学生对知识的系统把握和能力的提高。

本书可作为高等学校机械类、近机械类及其他工程类各专业开设的"工程图学"课程的教材，也可作为工程设计人员的参考用书。

本书由庄宏、叶福民主编。其中第5章、第6章、第9章由叶福民编写，其余各章由庄宏编写。

在本书编写过程中得到了江苏科技大学一流本科专业建设基金资助，在此表示诚挚的谢意。由于编者水平有限，书中难免存在不足之处，恳请读者批评指正。

庄宏

2023年2月

Contents 目 录

Chapter 1 Basic Knowledge of Drawing 第1章 制图基本知识 ···················· 1

Chapter 2 Projection of Geometric Elements 第2章 几何元素的投影 ·············· 6

Chapter 3 Solids and Their Intersections 第3章 立体及其交线 ···················· 19

Chapter 4 Composite Solids 第4章 组合体 ···················· 28

Chapter 5 General Principles of Representation 第5章 图样画法 ···················· 38

Chapter 6 Standard Parts and Commonly Used Parts 第6章 标准件与常用件 ···················· 50

Chapter 7 Detail Drawings 第7章 零件图 ···················· 59

Chapter 8 Assembly Drawings 第8章 装配图 ···················· 77

Chapter 9 Computer Drawing 第9章 计算机绘图 ···················· 90

References 参考文献 ···················· 105

1-1
参考答案

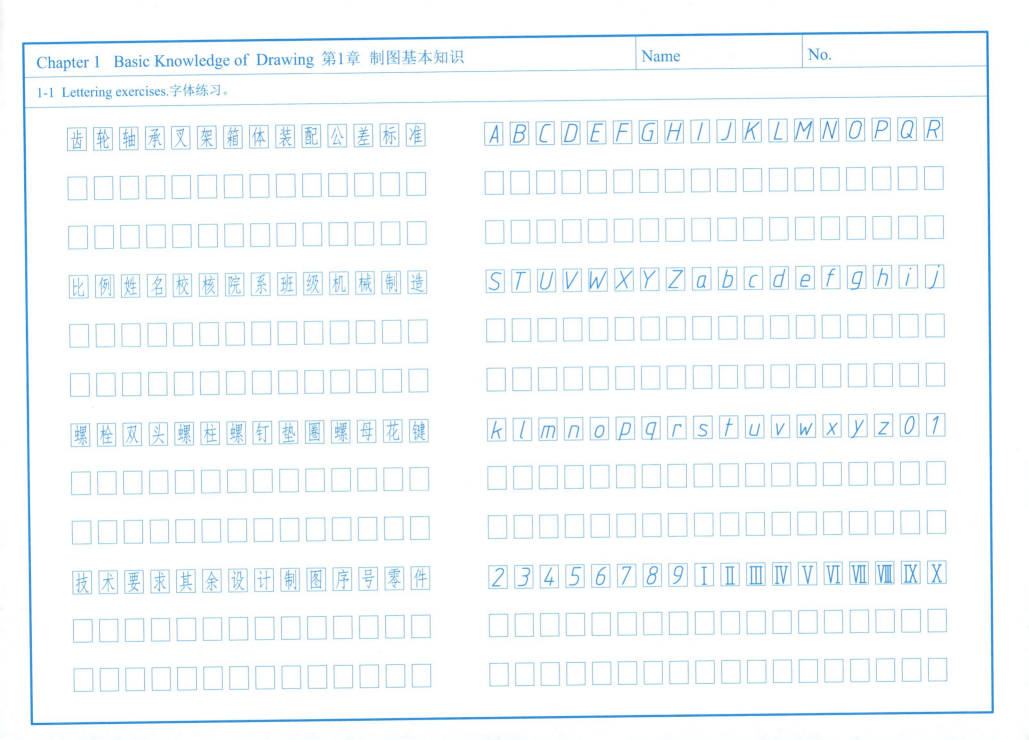

1-2
参考答案

| Chapter 1   Basic Knowledge of Drawing   第1章 制图基本知识 | Name | No. |

1-2 Refer to left pictures to complete the arc connection drawings and mark the centers of the connecting arcs. 参照左图完成圆弧连接作图，标出连接圆弧圆心。

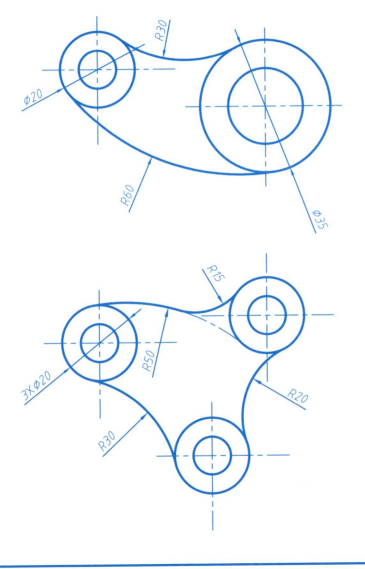
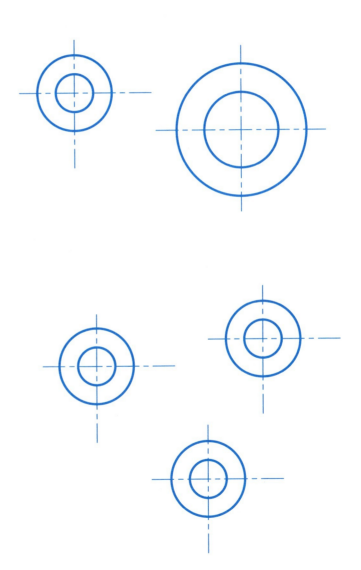

1-3
参考答案

| Chapter 1 Basic Knowledge of Drawing 第1章 制图基本知识 | Name | No. |

1-3 Redraw the plane figure and dimension following the example(on the scale of 1：1) at the appointed position . 根据图示尺寸，用1：1的比例在指定位置处画出平面图形并标注尺寸。

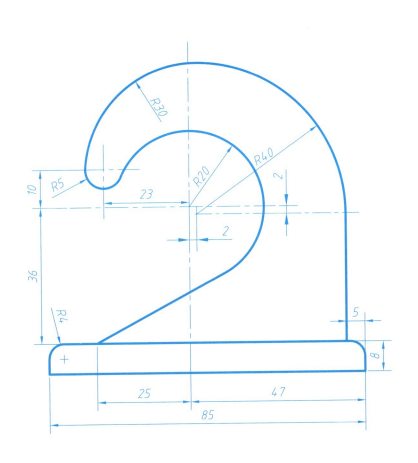

1-4
参考答案

| Chapter 1 Basic Knowledge of Drawing 第1章 制图基本知识 | Name | No. |

1-4 Redraw the plane figure and dimension following the example(on the scale of 1∶1) at the appointed position . 根据图示尺寸，用1∶1的比例在指定位置处画出平面图形并标注尺寸。

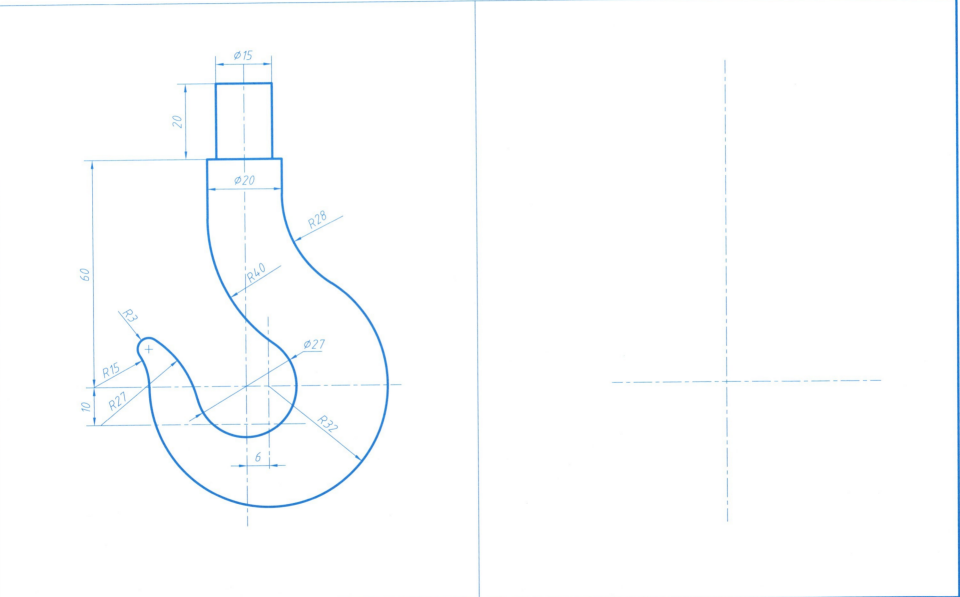

1-5
参考答案

| Chapter 1 Basic Knowledge of Drawing 第1章 制图基本知识 | Name | No. |

1-5 Redraw the plane figure and dimension following the example(on the scale of 1∶1) on an A3 sized sheet. 根据图示尺寸，用1∶1的比例在A3图纸上画出平面图形并标注尺寸。

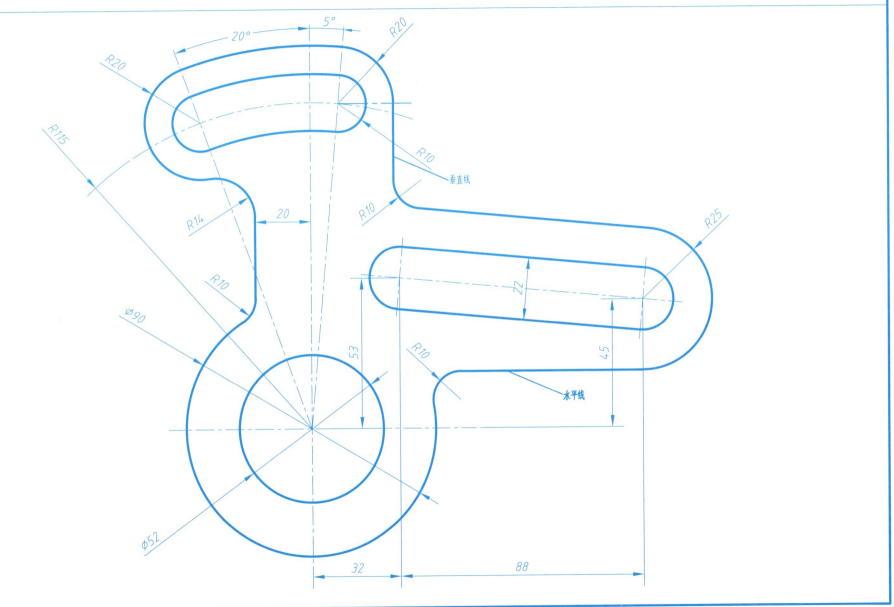

2-1
参考答案

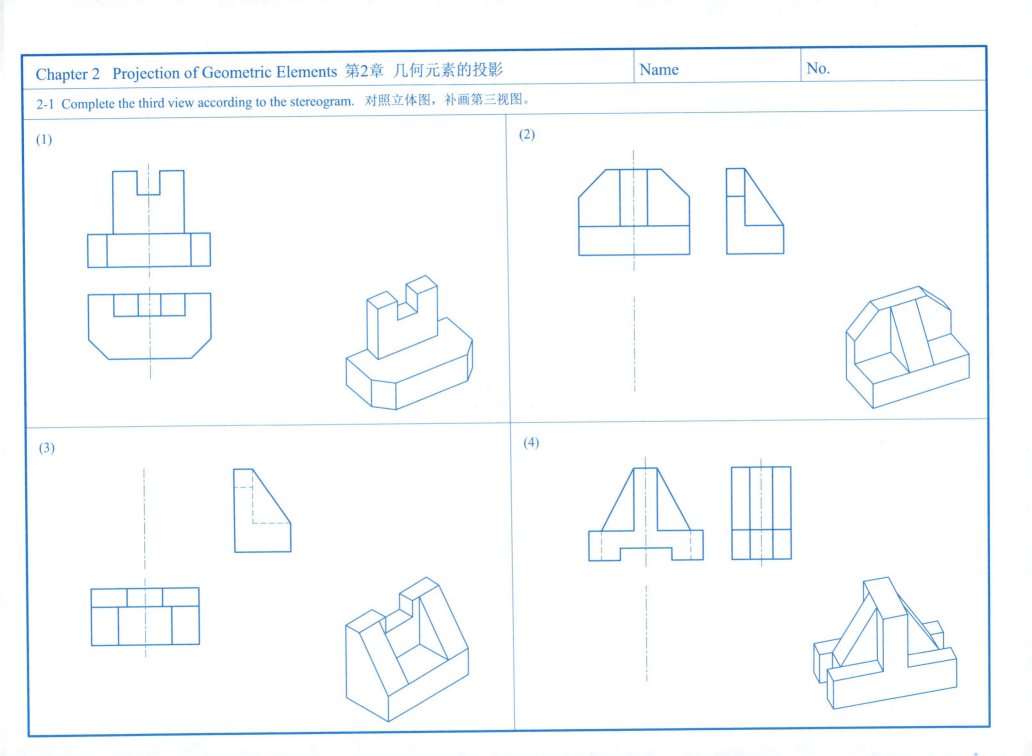

2-2到2-5
参考答案

Chapter 2 Projection of Geometric Elements 第2章 几何元素的投影

Name | No.

2-2 Complete the third view according to the others. 已知下列各点的两面投影，求作其他投影。

2-3 Draw the three projections of the given points A (20,12,0), B (0,20,18) and C (30,0,0). 已知A点(20,12,0)、B点(0,20,18)、C点(30,0,0)，求作各点投影。

2-4 Draw the three projections of each point given that the coordinate of point A is (20,10,15), point B is located 10 mm on the left side of A, 10 mm behind A and 15 mm above A, and point C is located directly in front of A. 已知A点(20,10,15)，B点在A点之左10 mm，在A点之前10 mm，在A点之上8 mm，C点在A点正前方15 mm，作出A、B和C的三面投影。

2-5 Draw the three projections of each point: (1) Point A is on plane V, 20 mm away from the X-axis, 30 mm away from plane W; (2) Point B is on the X-axis and is 20 mm away from plane W; Point C is on the Y-axis and is 18 mm away from plane V. 作出下列点的三面投影：(1) 点A在V面上，距X轴20 mm，距W面30 mm；(2) 点B在X轴上，距W面20 mm；(3) 点C在Y轴上，距V面18 mm。

2-6到2-9
参考答案

Chapter 2 Projection of Geometric Elements 第2章 几何元素的投影

| Name | No. |

2-6 Complete the three projections of line *AB* and *CD* according to point *A*(20,15,8), point *B*(5,15,18) and the two given views of line *CD*. 已知点*A*(20,15,8)，点*B*(5,15,18)及直线*CD*的两面投影，求作直线*AB*和*CD*的三面投影。

2-7 Decide the projection position of lines *SA*、*SB*、*AB* and *AC* according to the three views of the triangular pyramid. 已知三棱锥的三面投影，判断棱线*SA*、*SB*、*AB*和*AC*分别为何种位置直线。

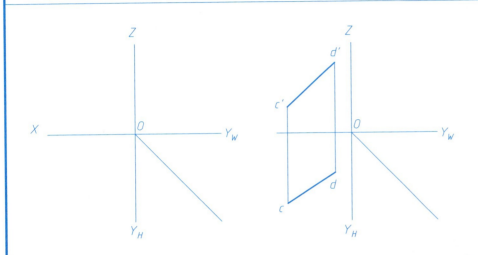
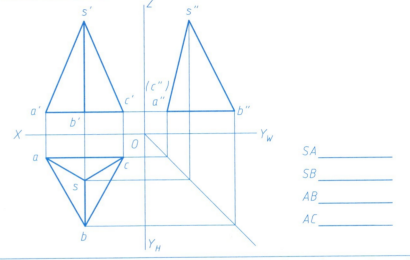

SA＿＿＿＿＿
SB＿＿＿＿＿
AB＿＿＿＿＿
AC＿＿＿＿＿

2-8 Draw a frontal line *AB* through point *A* and a *V*-perpendicular line *CD* through point *C* according to the three views of point *A* and point *C* respectively. The length of line *AB* is 15 mm, α=30°; the length of line *CD* is 12 mm. 已知点*A*、*C*的三面投影，过*A*点作一正平线*AB*，使其长为15 mm，α=30°，过点*C*作一正垂线*CD*，使其长为12 mm。

2-9 Complete the horizontal projection of point *C* according to the two views of line *AB* and the front projection of point *C* on the line. 已知直线*AB*的两面投影以及直线上点*C*的正面投影，求*C*点的水平投影。

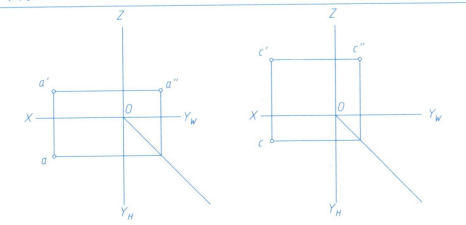
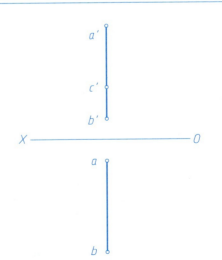

2-10到2-13
参考答案

| Chapter 2　Projection of Geometric Elements　第2章　几何元素的投影 | Name | No. |

2-10 Find the real length of each line, inclination α of line AB to plane H and inclination β of line CD to plane V. 作图求直线AB、CD的实长及AB对H面的倾角α、直线CD对V面的倾角β。

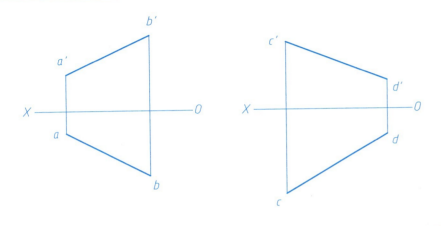

2-11 Complete the horizontal projection of line AC given that line AB and line AC have the same length. 已知直线AB和直线AC的长度(实长)相等，求作直线AC的水平投影。

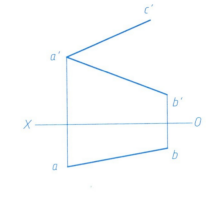

2-12 Draw the front projections of lines AB、CD given that the length of line AB is 30 mm and the inclination of line CD to plane H is 30°. 已知直线AB的实长为30 mm，直线CD对H面的倾角为30°，分别作出AB、CD的正面投影。

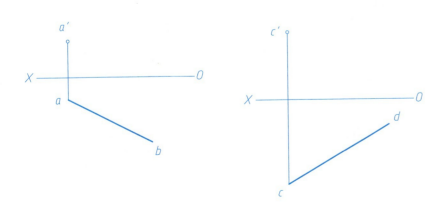

2-13 Draw point C on line AB such that the length of line BC is 18 mm. The distance between point C and plane H and plane V is _____ and _____ respectively. (accurate to 0.1 mm) 在直线AB上求作一点C，使得BC=18 mm。点C与H面和V面之间的距离分别为_____和_____。(精确到0.1 mm)

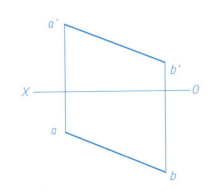

2-14到2-17
参考答案

| Chapter 2 Projection of Geometric Elements 第2章 几何元素的投影 | Name | No. |

2-14 Decide the relative positions of lines *AB*、*CD* and fill in the blanks. 判断直线 *AB* 与直线 *CD* 的相对位置，并将答案填写在横线上。

2-15 Draw the two views of line *EF* according to the projections of line *AB* and line *CD*; the horizontal line *EF* is 20 mm away from plane *H*, and intersects both line *AB* and line *CD*. 已知直线 *AB* 和直线 *CD* 的投影，另一水平线 *EF* 与 *H* 面的距离为20 mm，并且与 *AB*、*CD* 均相交，求作直线 *EF* 的两面投影。

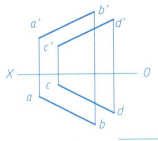
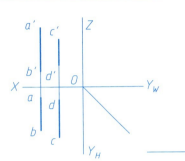
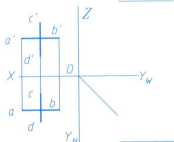
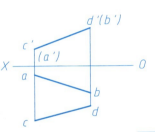

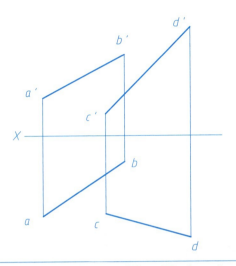

2-16 Decide the relative position of line *AB* and line *CD*, and mark the coincident points. 判断直线 *AB* 和直线 *CD* 的相对位置，并标注重影点的投影。

2-17 Draw the horizontal projection of line *CD* given that line *AB* is parallel to line *CD*. 已知直线 *AB* 与直线 *CD* 平行，求作 *CD* 的水平投影。

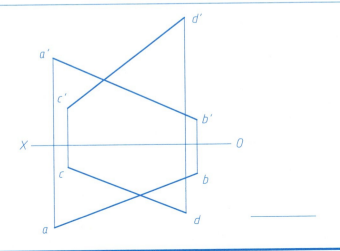
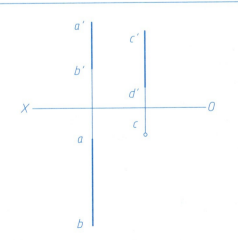

2-18到2-21
参考答案

Chapter 2 Projection of Geometric Elements 第2章 几何元素的投影	Name	No.

2-18 Draw the two views of the frontal line *KM* through point *K* given that line *KM* and line *AB* intersect at point *M*. 过已知点*K*作一正平线*KM*，与已知直线*AB*相交于点*M*。求出*KM*的两面投影。

2-19 Draw a line *MN* through point *M* given that line *MN* intersects the frontal line *AB* at point *K* perpendicularly. 过点*M*作直线*MN*与正平线*AB*垂直相交（正交）于点*K*。

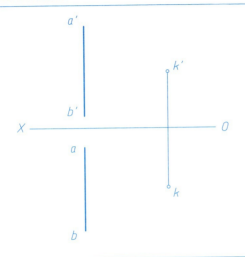

2-20 Draw the front projection of the isosceles triangle *ACB* given that the base *AB* is a horizontal line. 已知底边*AB*为水平线，求作等腰三角形*ACB*的正面投影。

2-21 Complete the two views of the rhombus *ABCD* according to the projection of the diagonal *BD* of the rhombus and the horizontal projection of the endpoint *A* of the other diagonal. 已知菱形*ABCD*的对角线*BD*的投影以及另一对角线端点*A*的水平投影，完成菱形的两面投影。

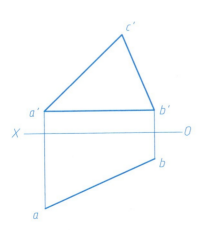

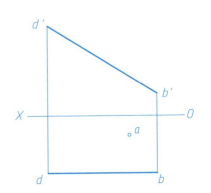

2-22到2-25
参考答案

Chapter 2 Projection of Geometric Elements 第2章 几何元素的投影

2-22 Decide the projection type of the following planes. 判别下列平面分别是什么位置平面。

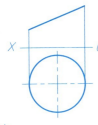

(1) _____

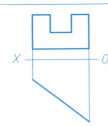

(2) _____

(3) _____

(4) _____

2-23 Complete the horizontal projection of the pentagon ABCDE. 完成平面五边形ABCDE的水平投影。

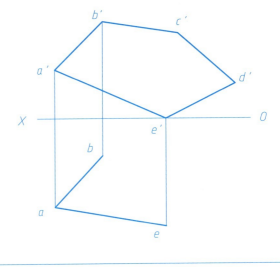

2-24 Complete the third view and fill in the blanks. 完成平面的侧面投影并填空。

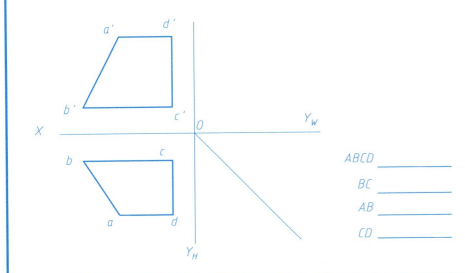

ABCD _____

BC _____

AB _____

CD _____

2-25 Draw the two views of the following planes through the specified lines (represented by a trace). 包含下列直线作平面(用迹线表示)。

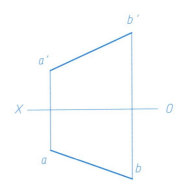

(1) V-perpendicular plane P
正垂面P

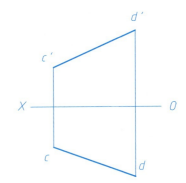

(2) H-perpendicular plane Q
铅垂面Q

2-26到2-29
参考答案

Chapter 2 Projection of Geometric Elements 第2章 几何元素的投影

Name | No.

2-26 Draw a *H*-perpendicular plane through line *AB*(represented by a triangle), draw a frontal plane through line *CD*(represented by an equilateral triangle). 包含直线*AB*作铅垂面(用三角形表示)，包含直线*CD*作正平面(用等边三角形表示)。

2-27 Draw the two views of the square given that the square *ABCD* is a *V*-perpendicular plane, and its diagonal *AC* is a frontal line. 已知正方形*ABCD*为正垂面，其对角线*AC*为正平线，作出该正方形的投影。

2-28 Draw a frontal line through point *K* on triangle *ABC*, and draw the front projection of point *K*. 在三角形*ABC*上过*K*点作正平线，并作出*K*点的正面投影。

2-29 Draw the other view of point *K* given that it is on the plane *ABCD*. 已知点*K*是平面*ABCD*上的一点，求其另一面投影。

2-30到2-32
参考答案

| Chapter 2   Projection of Geometric Elements 第2章 几何元素的投影 | Name | No. |

2-30 Draw a front line belongs to the △ABC plane given that the line is 35 mm in front of plane V, draw a horizontal line 30mm above plane H. 在△ABC平面上作正平线，该线在V面之前35 mm；作△ABC平面的水平线，该线在H面之上30 mm。

2-31 Complete the two views of plane ABCD given that it is a rectangle. 已知平面ABCD为矩形，完成其两面投影。

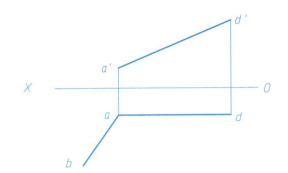

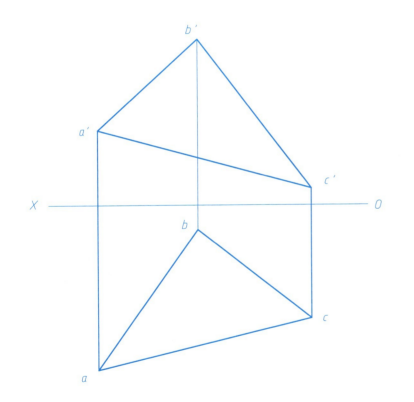

2-32 Decide wether points A、B、C and D are in the same plane. 判断点A、B、C、D是否在同一平面内。

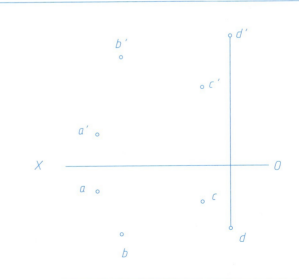

2-33到2-35
参考答案

| Chapter 2 Projection of Geometric Elements 第2章 几何元素的投影 | Name | No. |

2-33 Draw a plane (represented by a triangle) parallel to the known plane ABCD through point K. 过点K作平面(用三角形表示)平行于已知平面ABCD。

2-34 Draw line KF parallel to plane H and plane ABC. 作直线KF平行于H面和平面ABC。

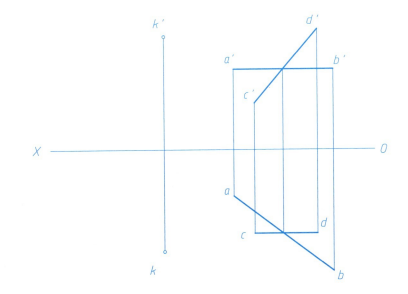

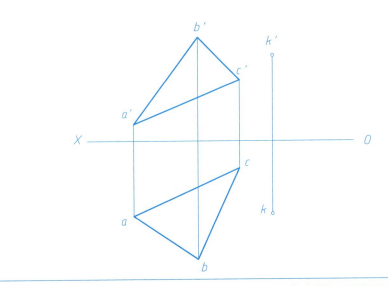

2-35 Draw line KF parallel to plane V and plane P. 作直线KF平行于V面和平面P。

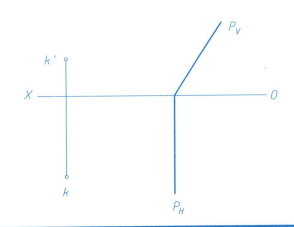

2-36到2-37
参考答案

| Chapter 2 Projection of Geometric Elements 第2章 几何元素的投影 | Name | No. |

2-36 Complete the front view of plane *GEF* given that plane *GEF* formed by the two parallel lines *AB* and *CD* is parallel to plane *GEF*. 已知由平行的两直线*AB*和*CD*构成的平面平行于平面*GEF*，完成平面*GEF*的正面投影。

2-37 Judge whether plane *ABC* and plane *DEF* are parallel to each other by drawing. 通过作图判断平面*ABC*和平面*DEF*是否相互平行。

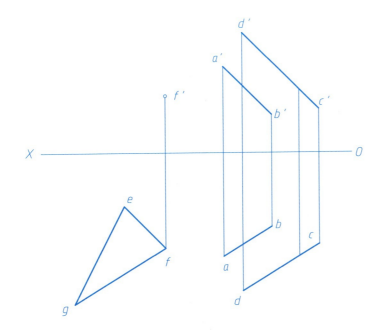

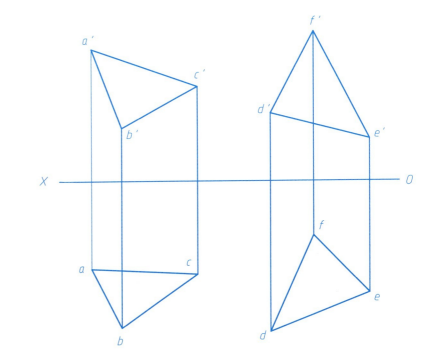

2-38
参考答案

Chapter 2 Projection of Geometric Elements 第2章 几何元素的投影

2-38 Draw the projections of the intersection K, and identify the visibility of the lines. 作出交点K的投影，并判别直线的可见性。

(1)

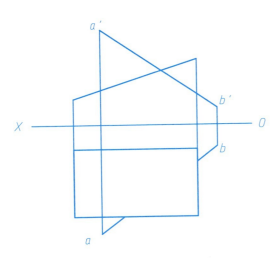

(2)

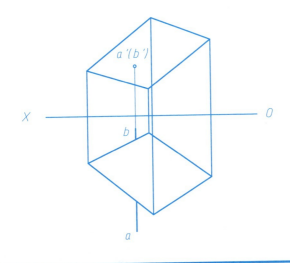

(3)

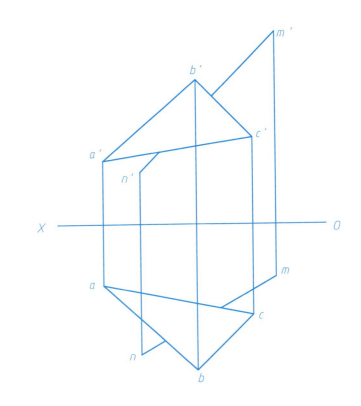

2-39
参考答案

| Chapter 2 Projection of Geometric Elements 第2章 几何元素的投影 | Name | No. |

2-39 Draw the projections of the intersection line of two planes, and identify their visibility. 作出两平面交线的投影，并判别其可见性。

(1)

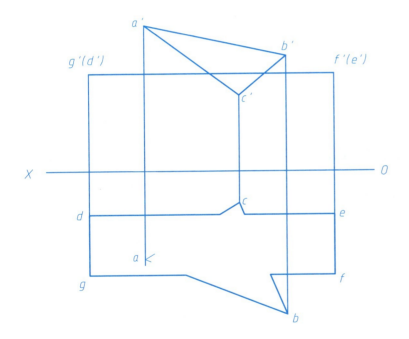

(2)

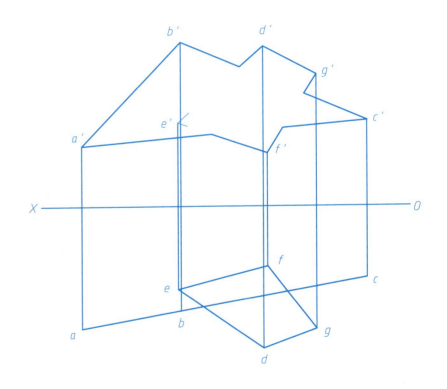

3-1(1-4)
参考答案

3-1(1-4)
动画／模型

| Chapter 3　Solids and Their Intersections　第3章 立体及其交线 | Name | No. |

3-1 Draw the third view of the solids and the projections of all points and lines, and identity their visibility. 画出立体的第三面投影，求立体表面上各点、线的其余两面投影，并判别其可见性。

(1)

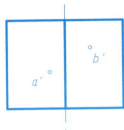

(2)

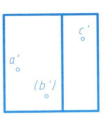

(3)

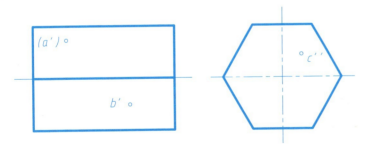

(4)

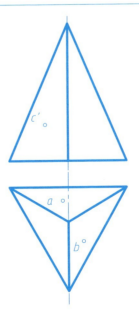

3-1(5-8)
参考答案

3-1(5-8)
动画／模型

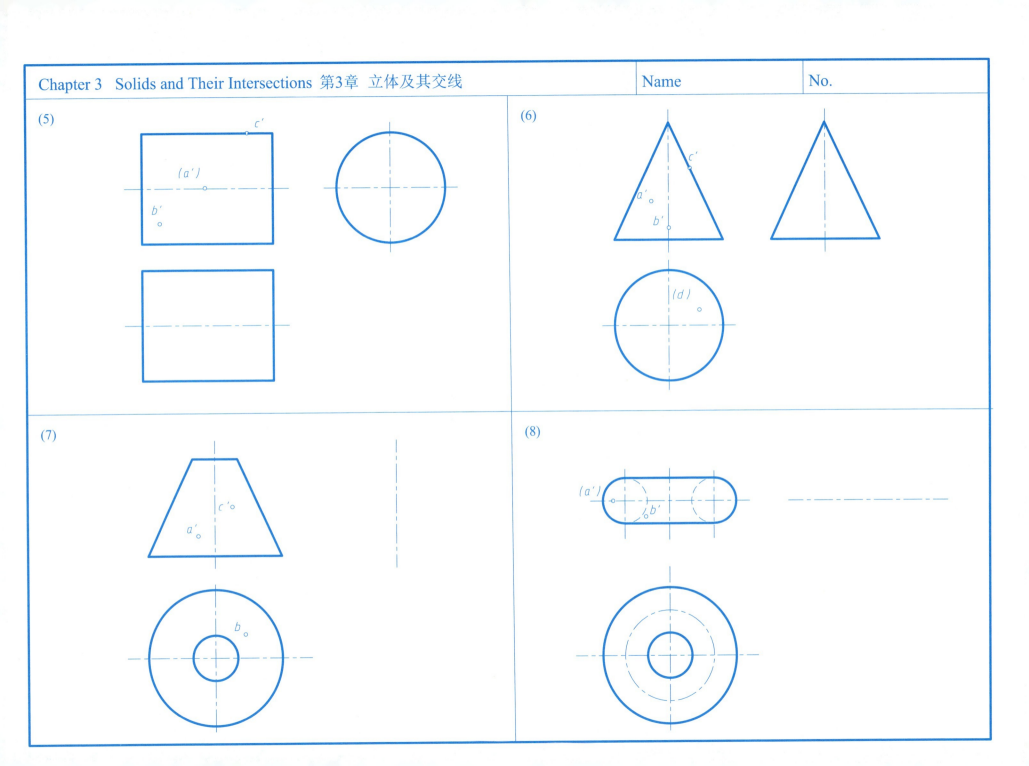

3-1(9-12)
参考答案

3-1(9-12)
动画 / 模型

Chapter 3 Solids and Their Intersections   第3章 立体及其交线   Name   No.

(9)

(10)

(11)

(12)

3-2(1-4)
参考答案

3-2(1-4)
动画／模型

Chapter 3 Solids and Their Intersections 第3章 立体及其交线

3-2 Complete the three views of the truncated solids. 完成截切立体的三面投影。

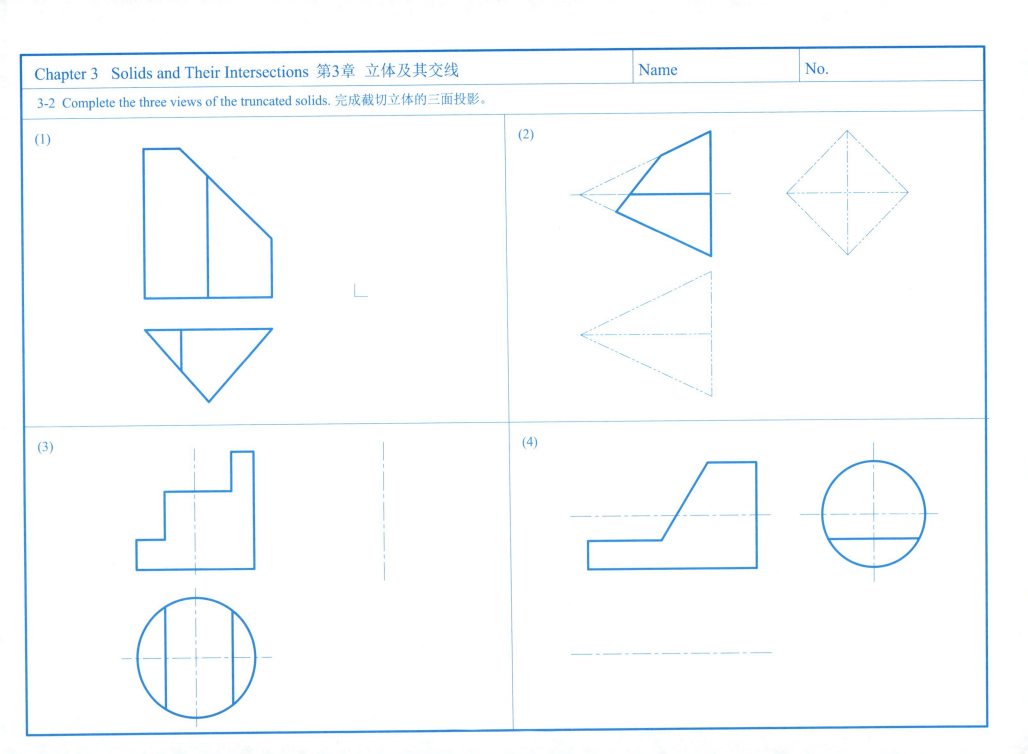

3-2(5-8)
参考答案

3-2(5-8)
动画/模型

| Chapter 3 Solids and Their Intersections 第3章 立体及其交线 | Name | No. |

(5)

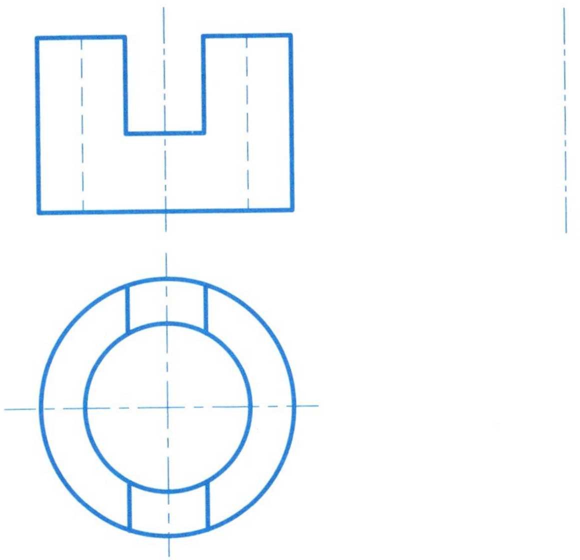

(6)

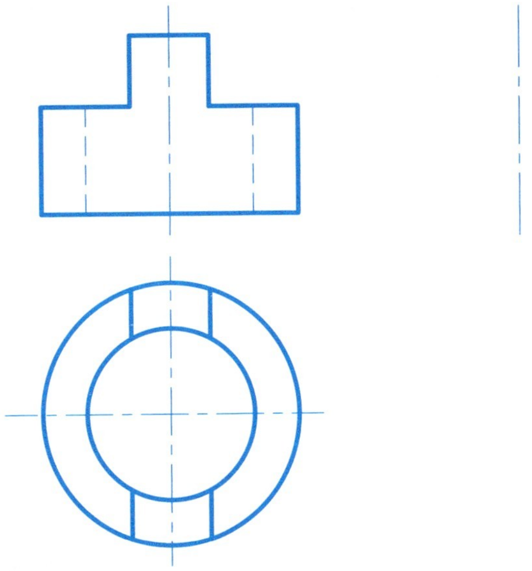

(7)

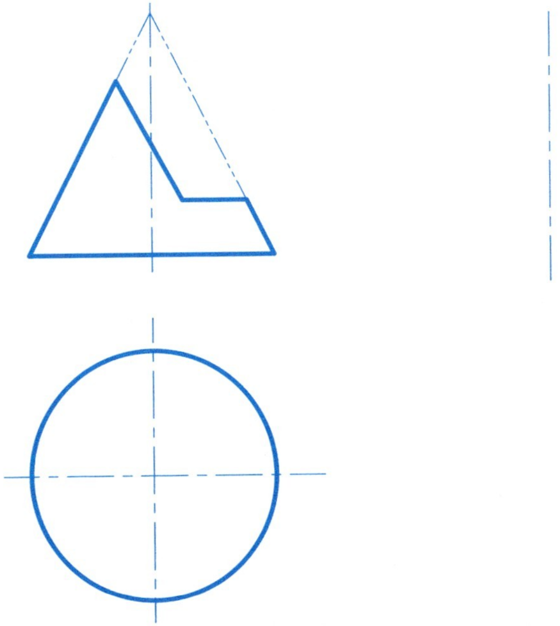

(8)

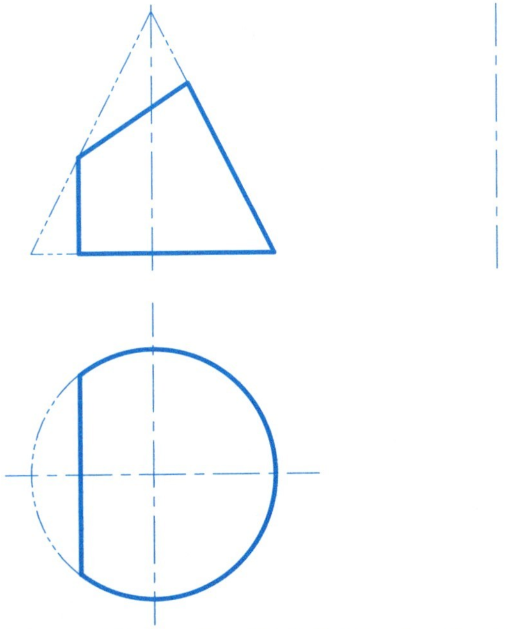

3-2(9-12)
参考答案

3-2(9-12)
动画／模型

| Chapter 3 Solids and Their Intersections 第3章 立体及其交线 | Name | No. |

(9)

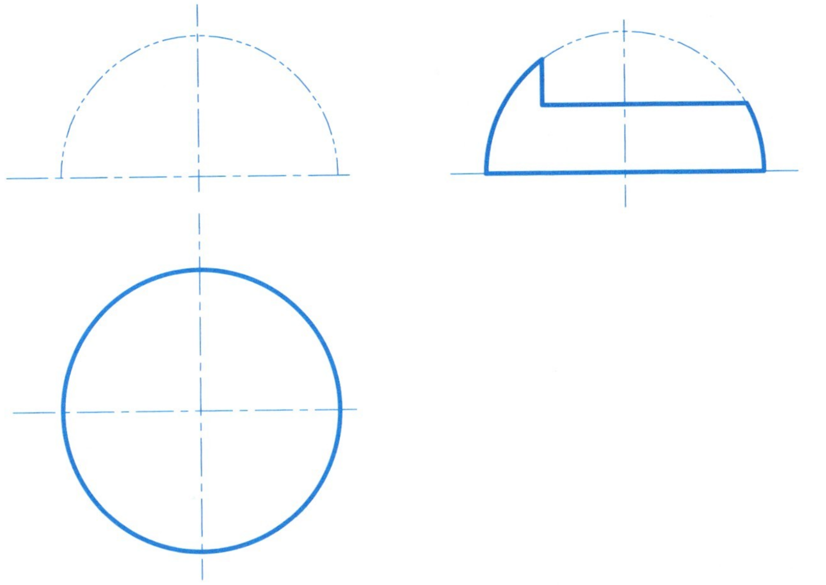

(10)

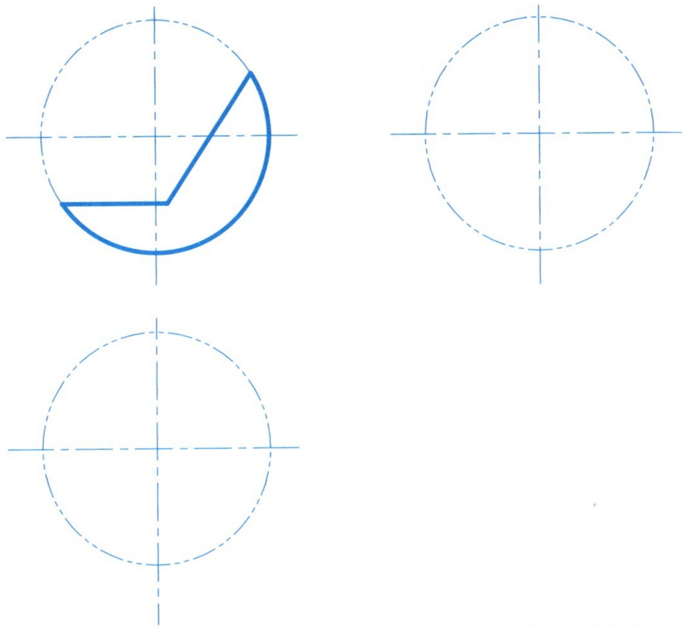

(11)

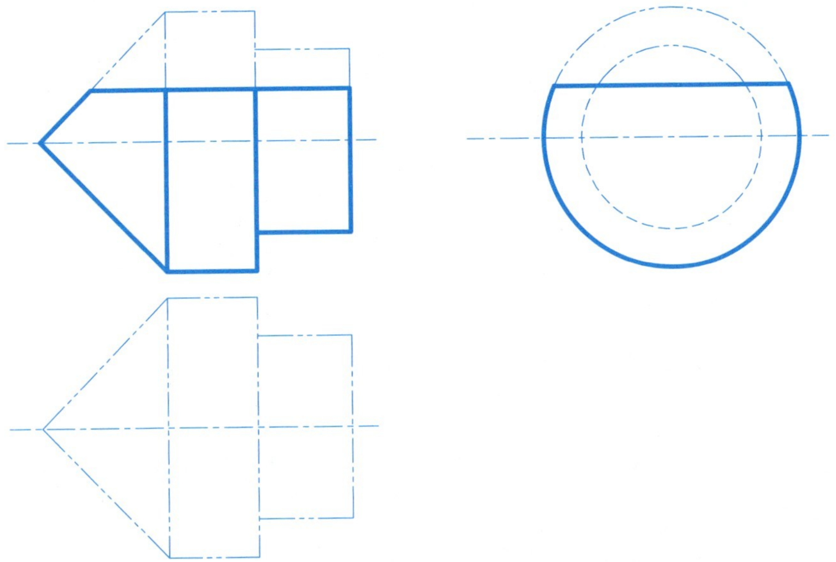

(12)

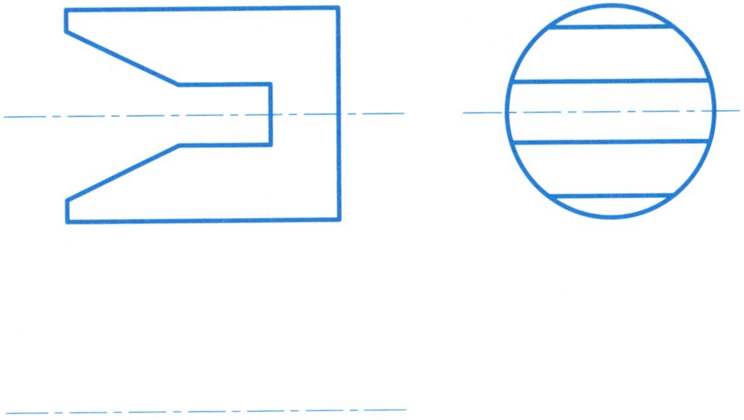

3-3(1-4)
参考答案

3-3(1-4)
动画／模型

| Chapter 3 Solids and Their Intersections 第3章 立体及其交线 | Name | No. |

3-3 Complete the three views of the intersecting bodies. 完成相贯立体的三面投影。

(1)

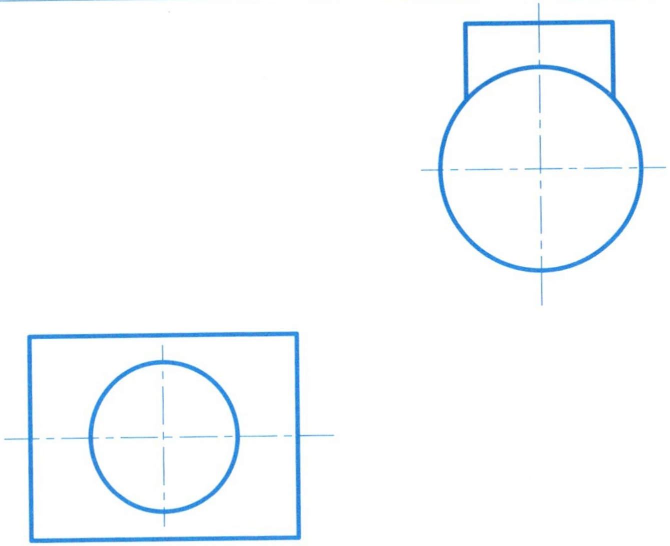

(2)

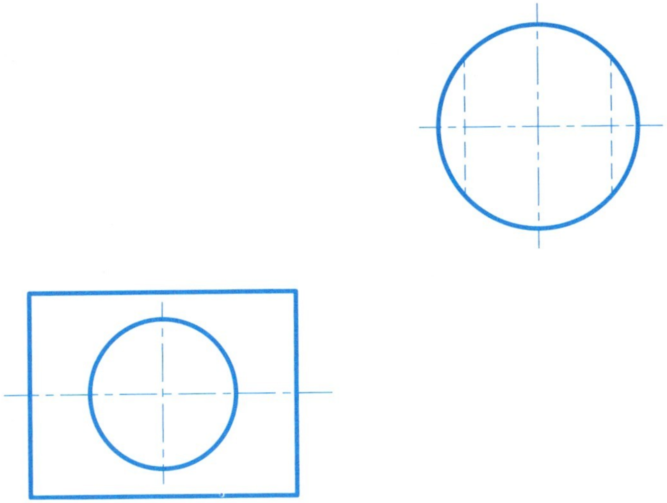

(3)

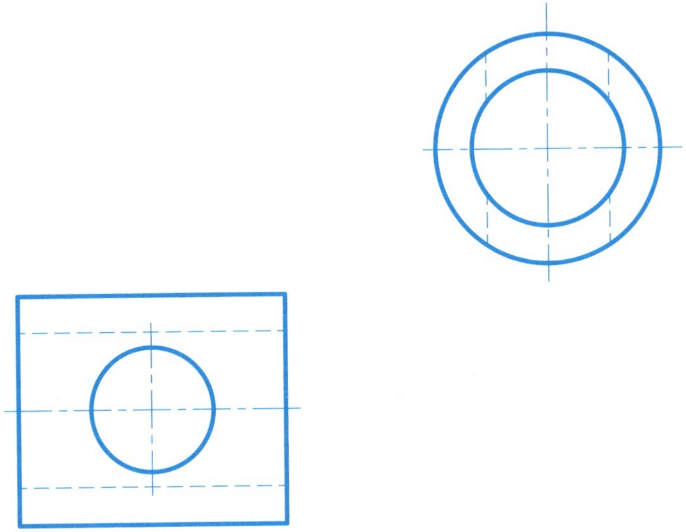

(4)

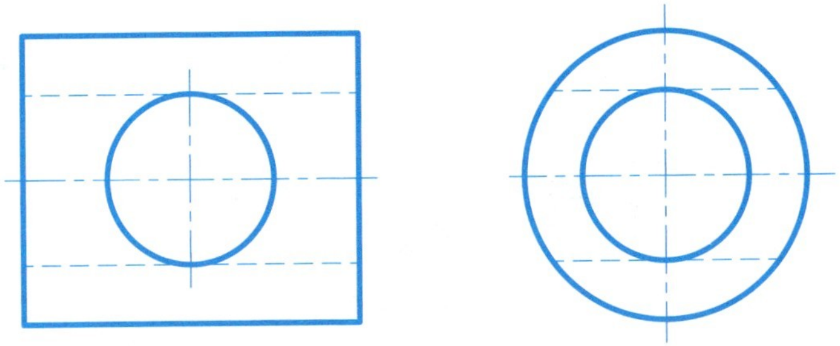

3-3(5-8)
参考答案

3-3(5-8)
动画／模型

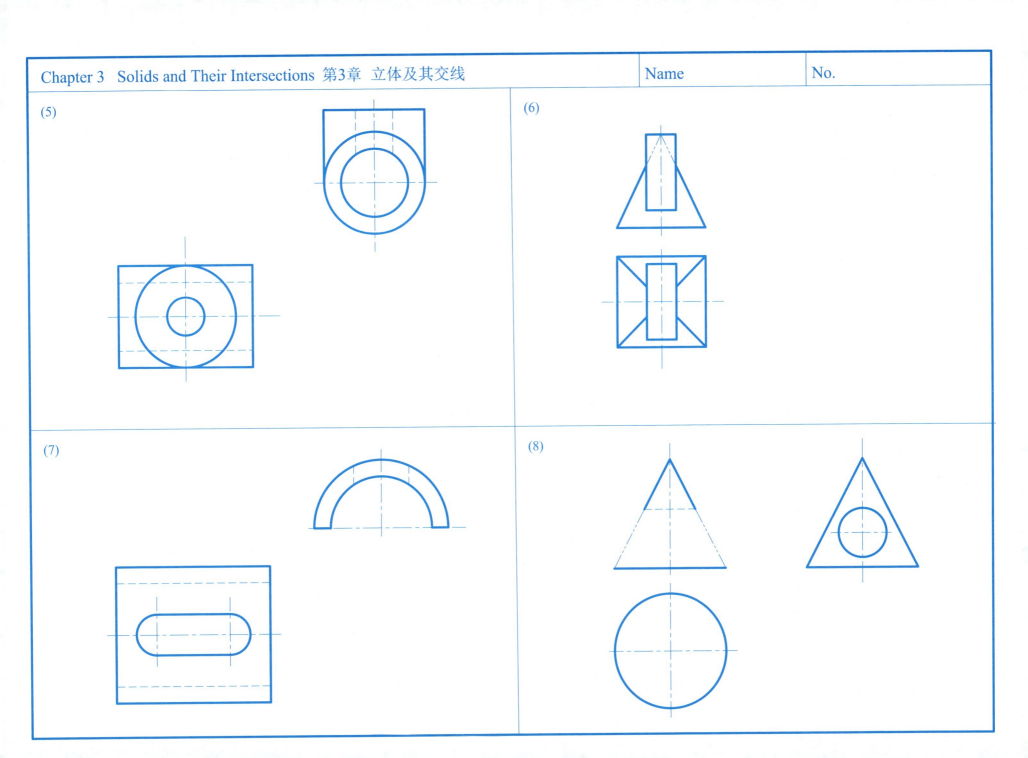

3-3(9-12)
参考答案

3-3(9-12)
动画／模型

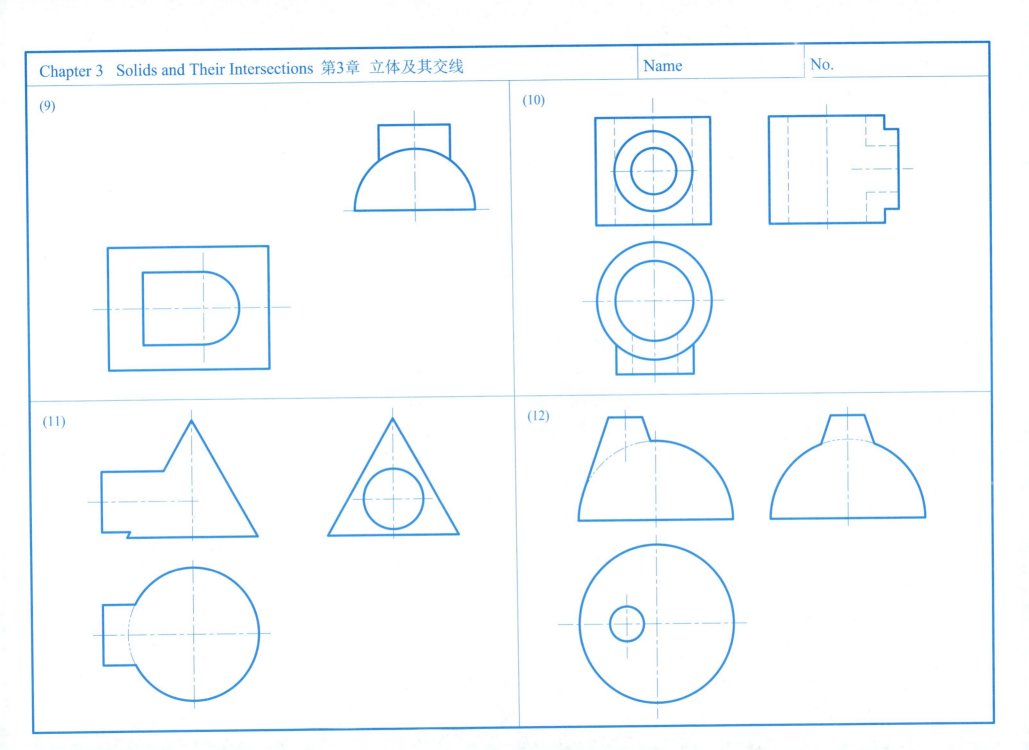

4-1(1-3)
参考答案

4-1(1-3)
动画／模型

| Chapter 4　Composite Solids　第4章　组合体 | Name | No. |

4-1　Draw the dimensions (take sizes from the drawings and hold the integer values). 标注尺寸(尺寸数值由图中测量，取整数)。

(1)

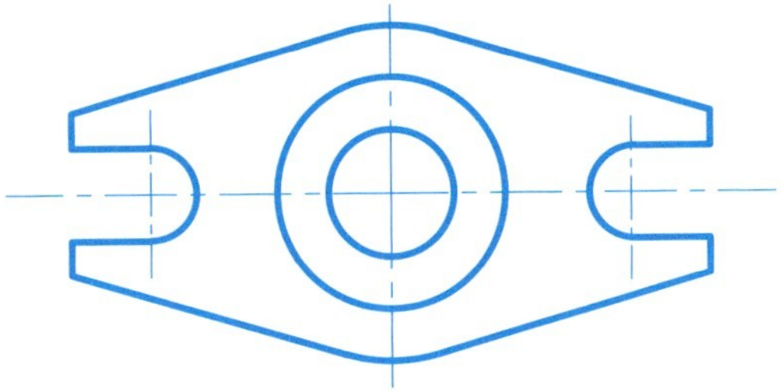

(2)

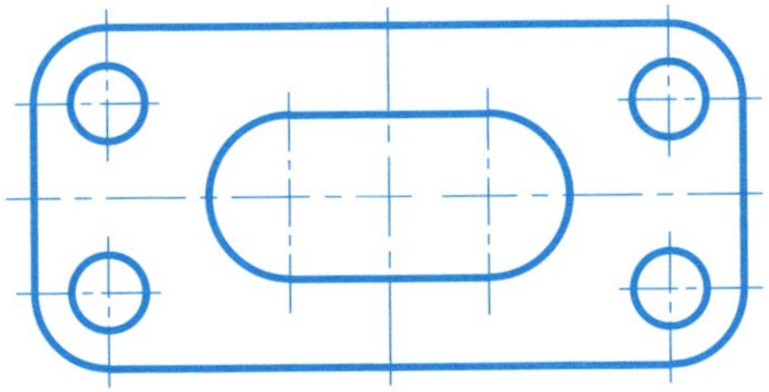

(3)

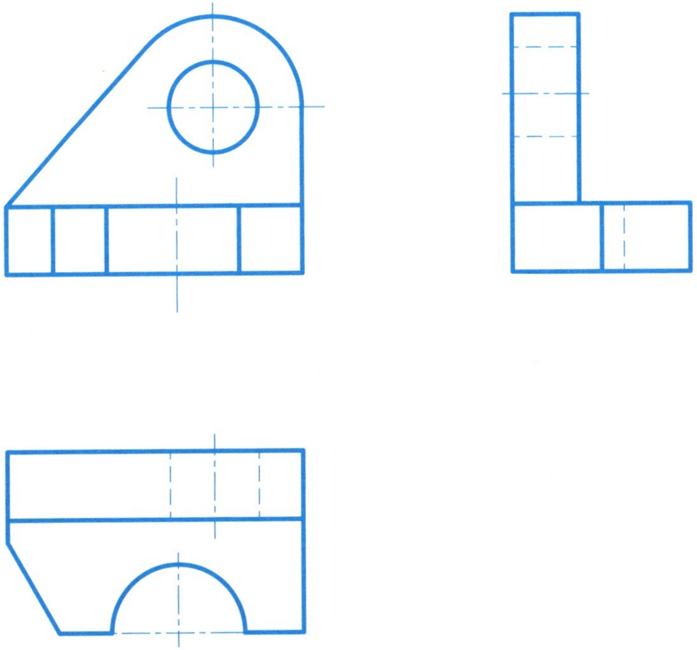

4-1(4-5)
参考答案

4-1(4-5)
动画／模型

| Chapter 4 Composite Solids 第4章 组合体 | Name | No. |

(4)

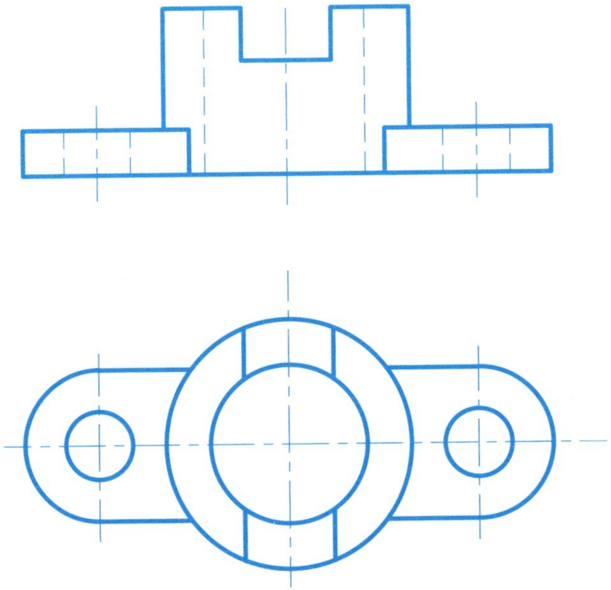

(5)

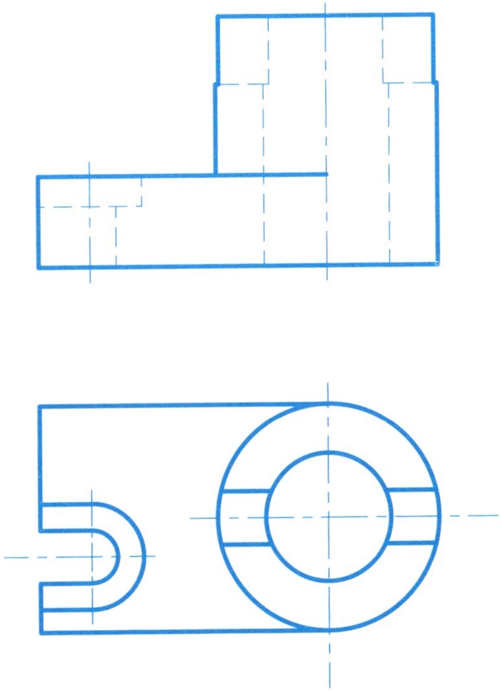

4-2(1-4)
参考答案

4-2(1-4)
动画 / 模型

| Chapter 4 Composite Solids 第4章 组合体 | Name | No. |

4-2 Draw the third view. 求作第三面投影。

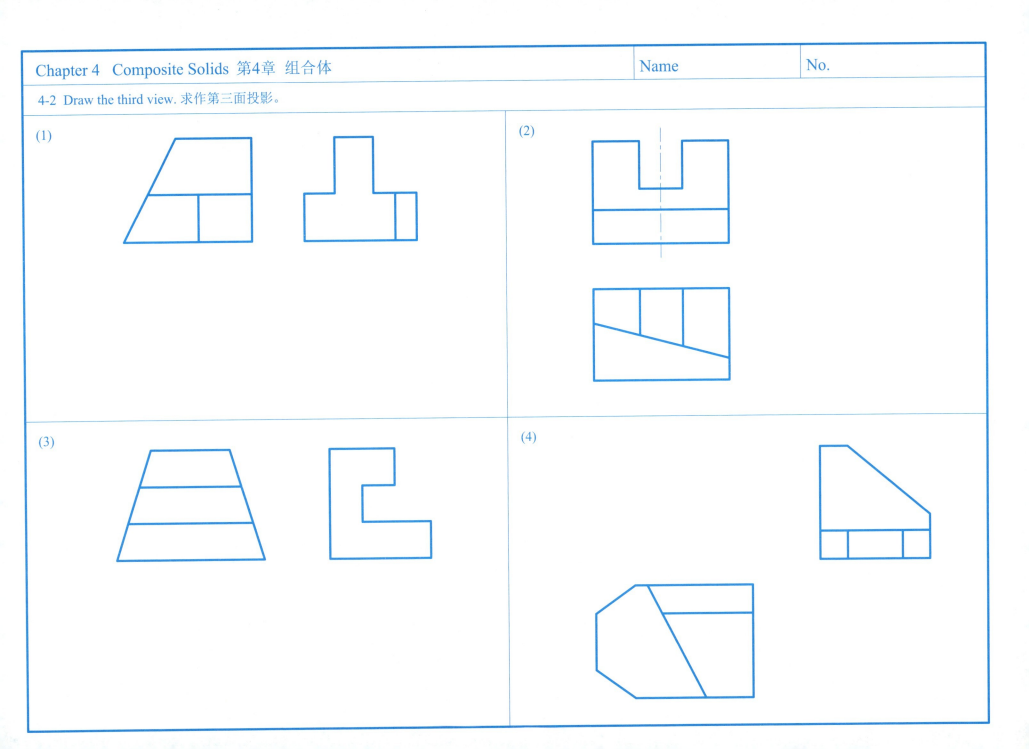

4-2(5-8)
参考答案

4-2(5-8)
动画 / 模型

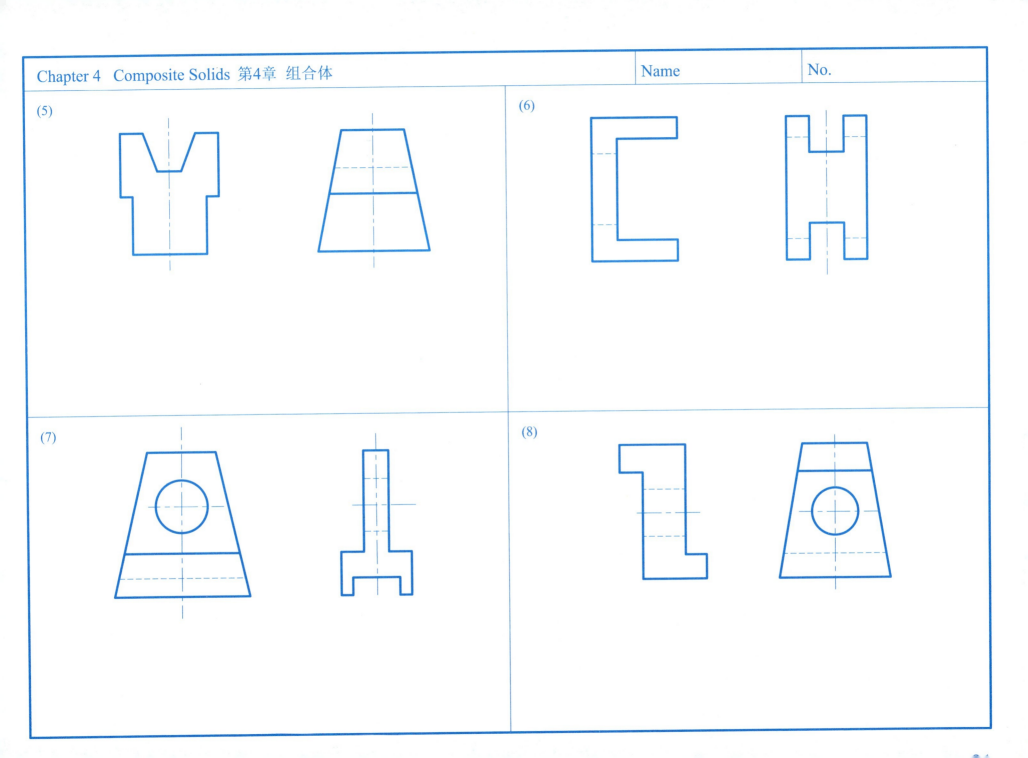

4-2(9-12)
参考答案

4-2(9-12)
动画／模型

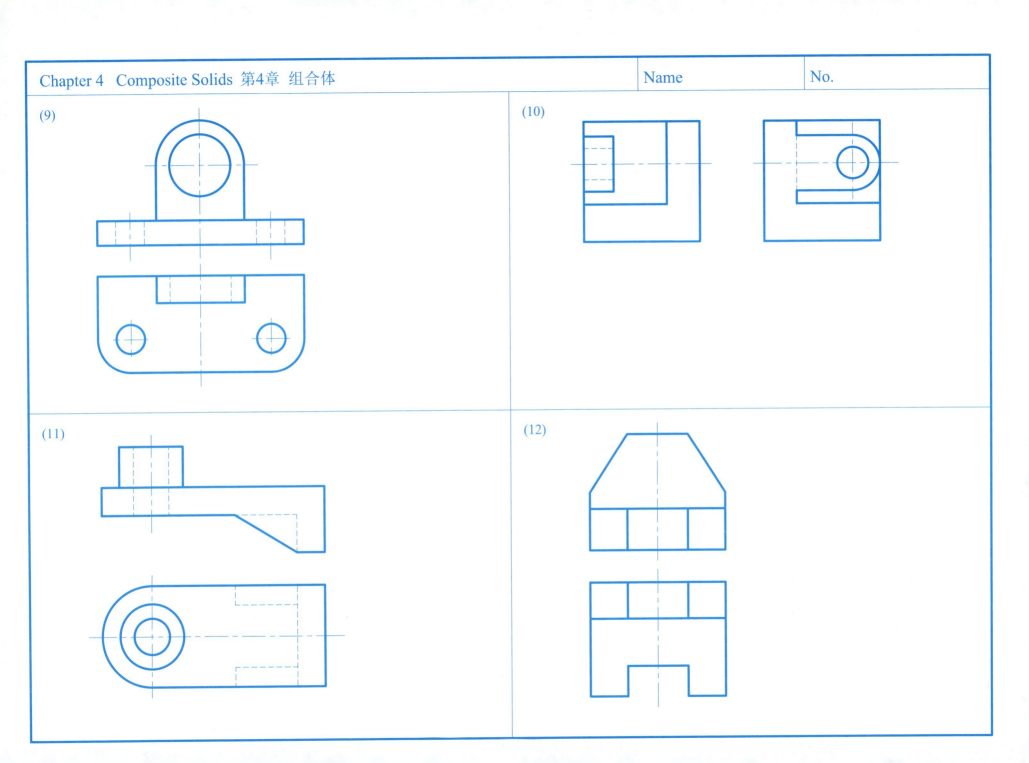

4-2(13-16)
参考答案

4-2(13-16)
动画／模型

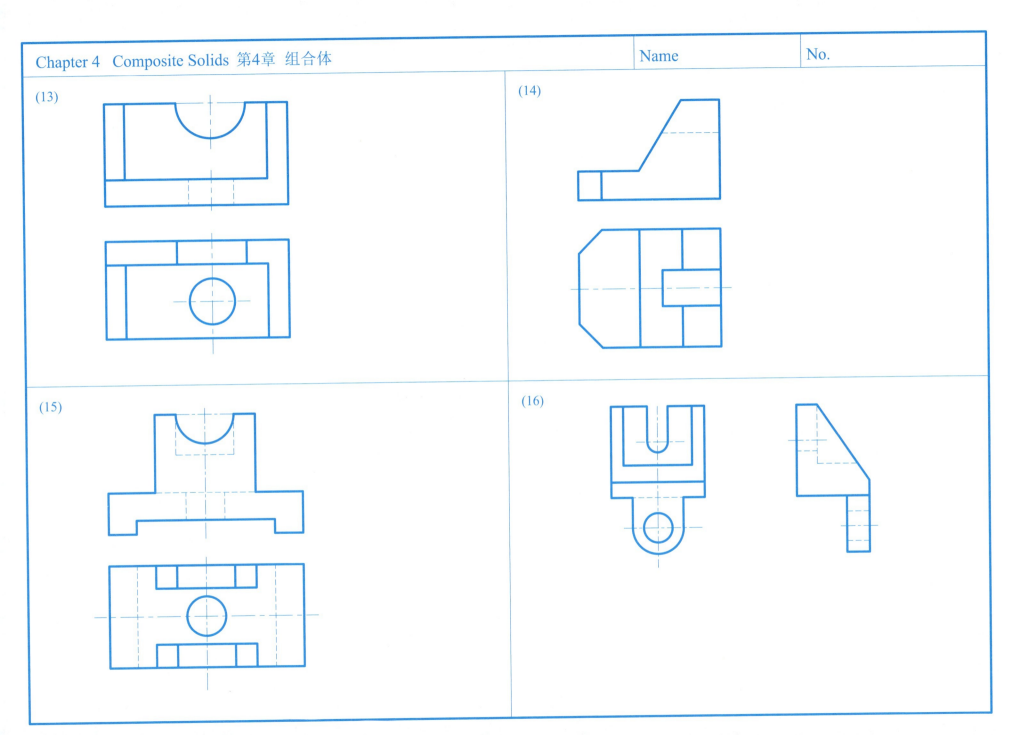

4-2(17-20)
参考答案

4-2(17-20)
动画／模型

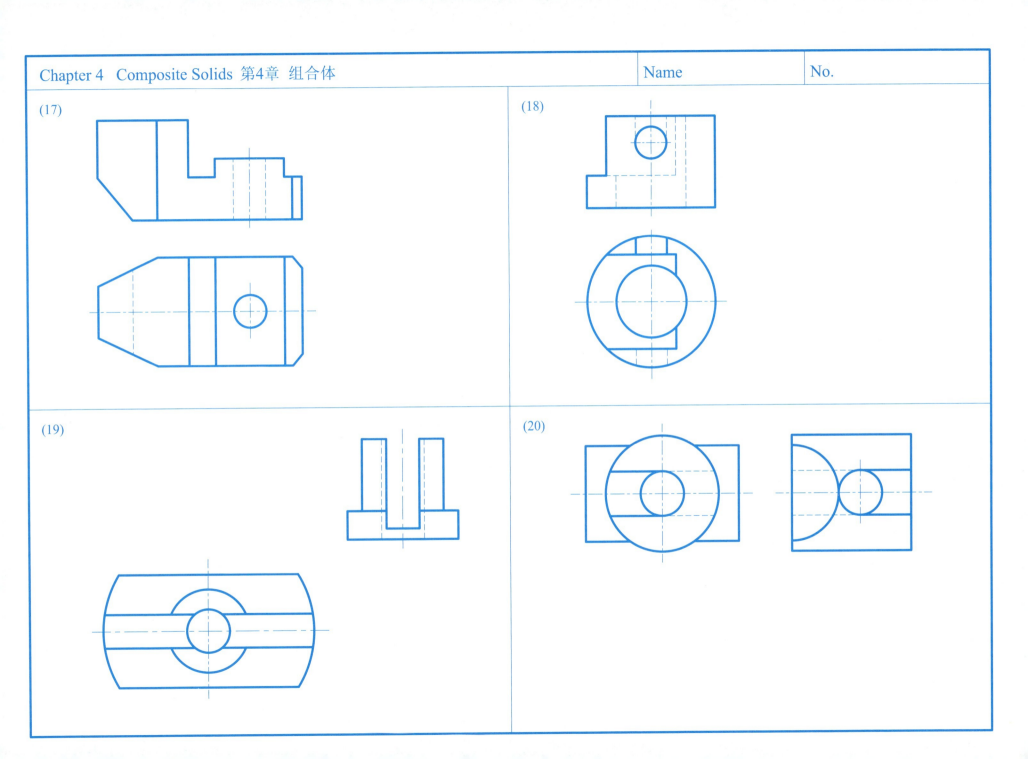

4-2(21-24)
参考答案

4-2(21-24)
动画／模型

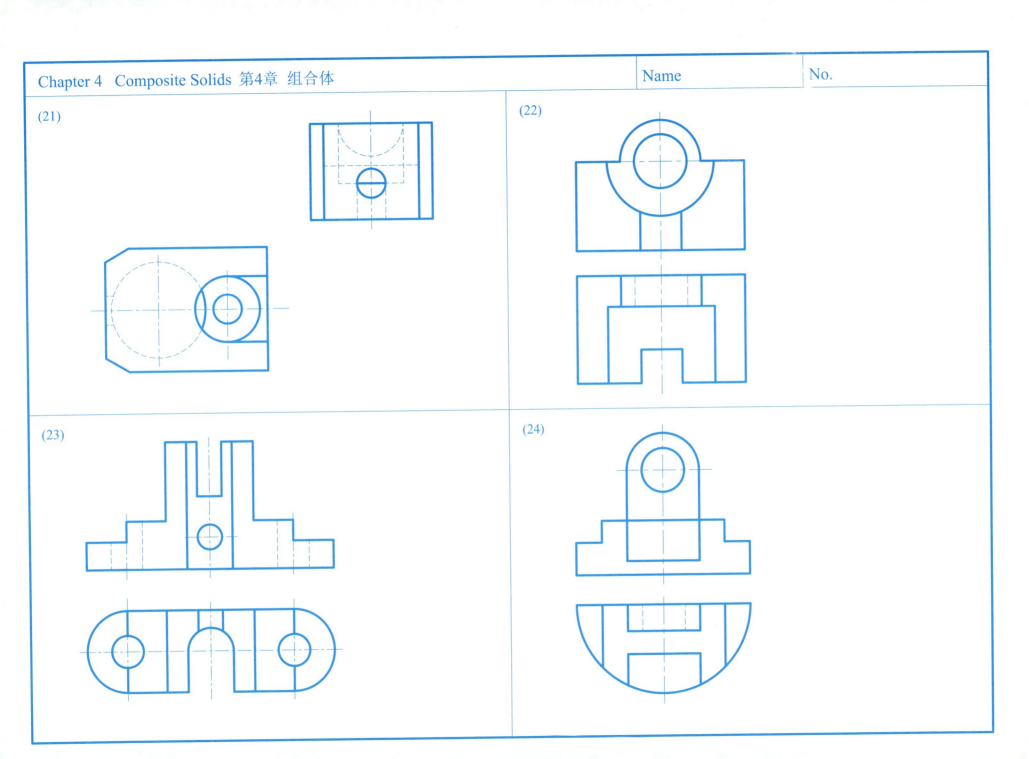

4-3(1-2)
参考答案

4-3(1-2)
动画／模型

| Chapter 4　Composite Solids　第4章　组合体 | Name | No. |

4-3　Draw the three views of the composite solids according to the given sizes（on the scale of 1∶1）. 由立体图画组合体三视图(按图中尺寸，用1∶1的比例画图)。

(1)

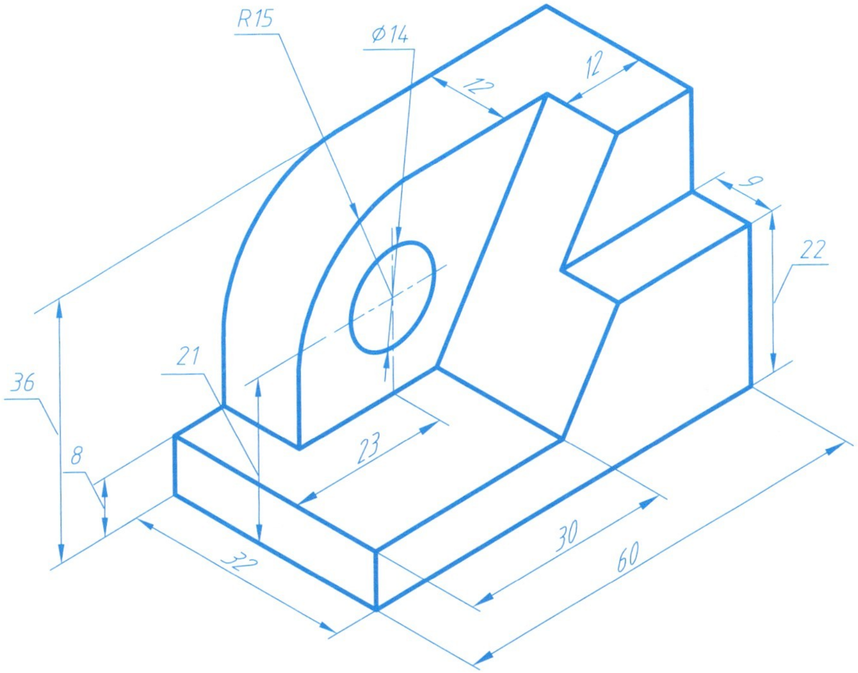

(2)

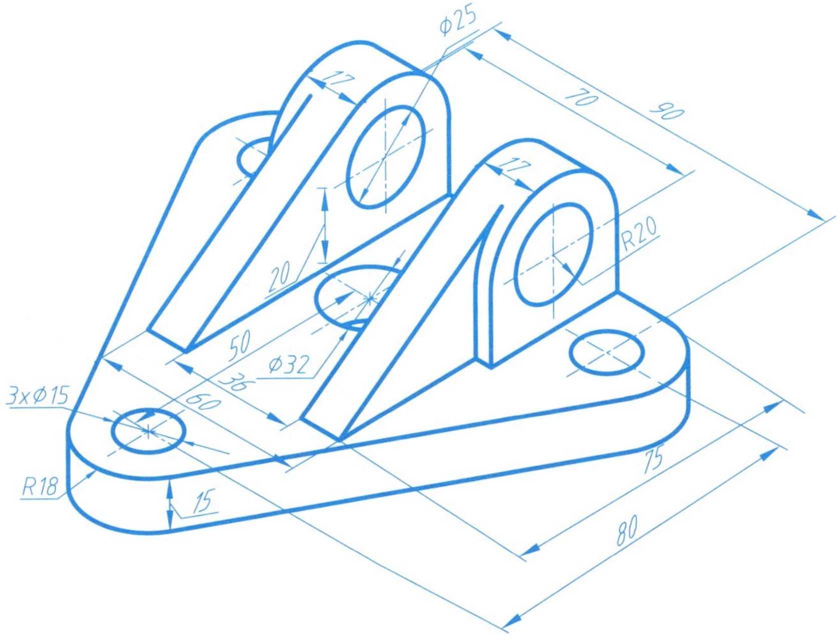

4-3(3-4)
参考答案

4-3(3-4)
动画／模型

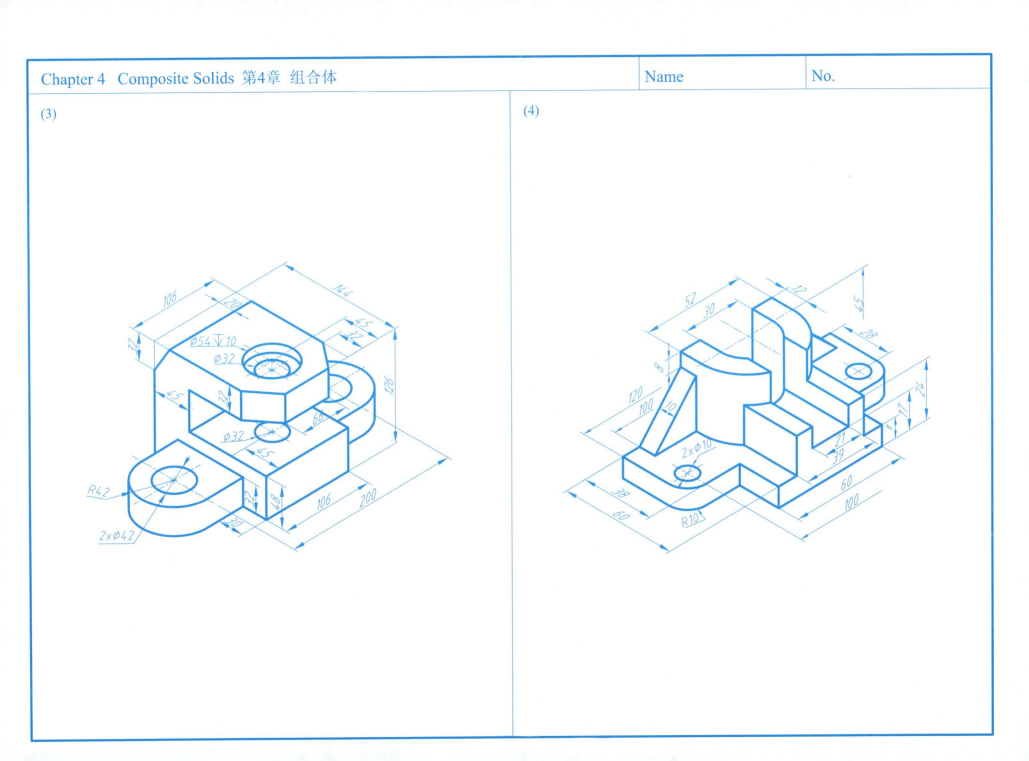

5-1到5-2
参考答案

5-1到5-2
动画／模型

Chapter 5 General Principles of Representation 第5章 图样画法	Name	No.
5-1 Draw views from *A* and *B* respectively. 作A向和B向视图。	5-2 Draw a inclined view from *A* and a partial view from *B*. 作A向斜视图和B向局部视图。	

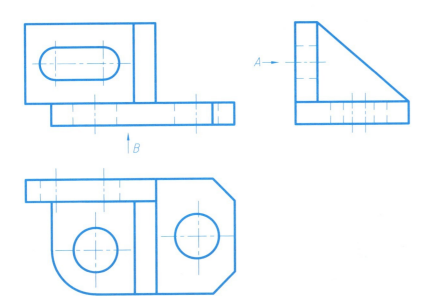

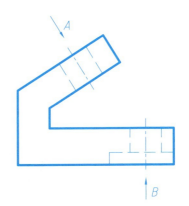

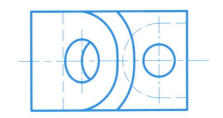

5-3(1-2)
参考答案

5-3(1-2)
动画／模型

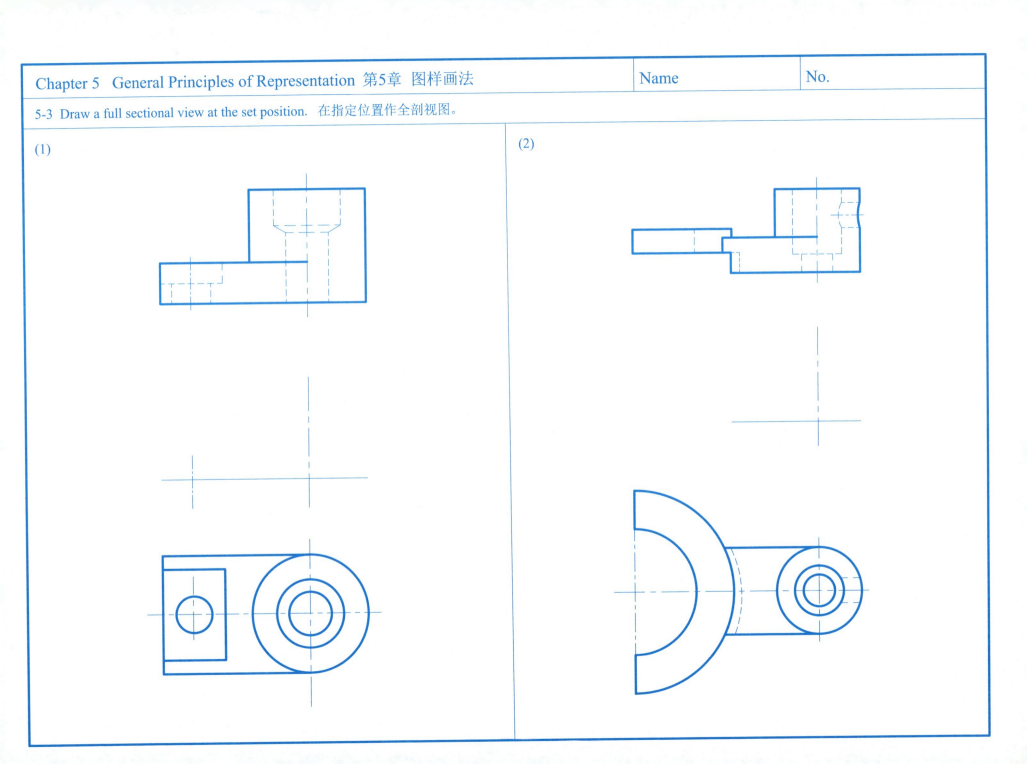

5-4(1-2)
参考答案

5-4(1-2)
动画／模型

| Chapter 5 General Principles of Representation 第5章 图样画法 | Name | No. |

5-4 Draw full sectional left-side views. 作出全剖的左视图。

(1)

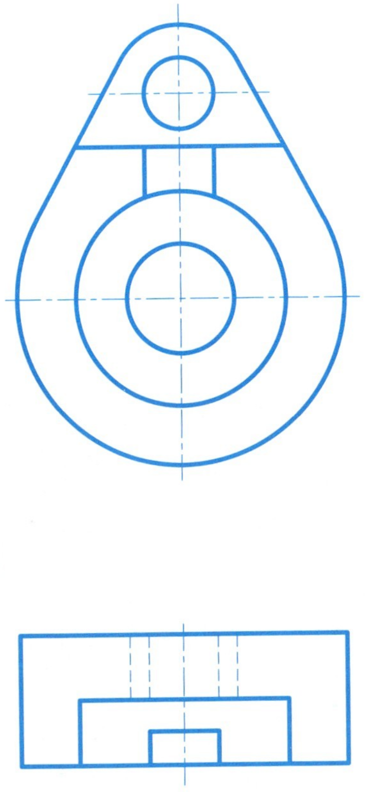

(2)

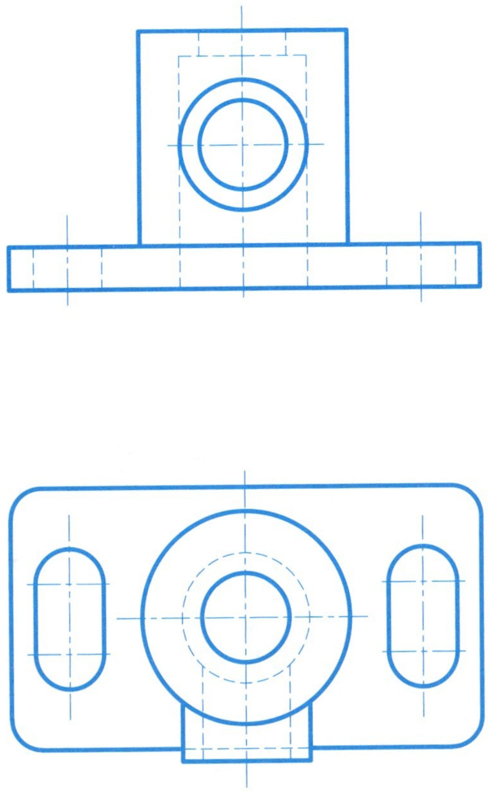

5-5(1-2)
参考答案

5-5(1-2)
动画／模型

| Chapter 5 General Principles of Representation 第5章 图样画法 | Name | No. |

5-5 Draw half sectional left-side views. 作出半剖的左视图。

(1)

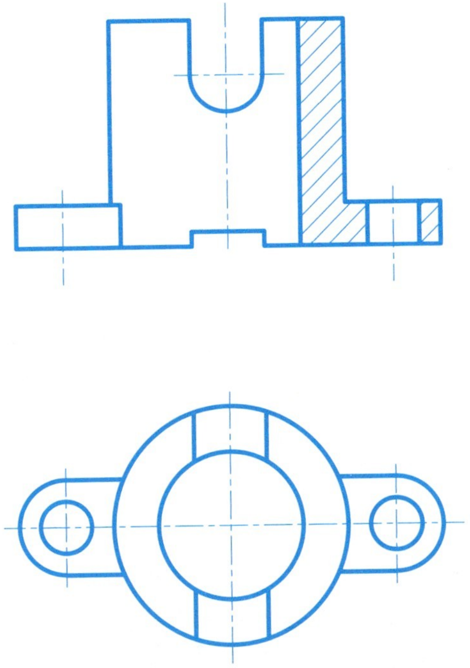

(2)

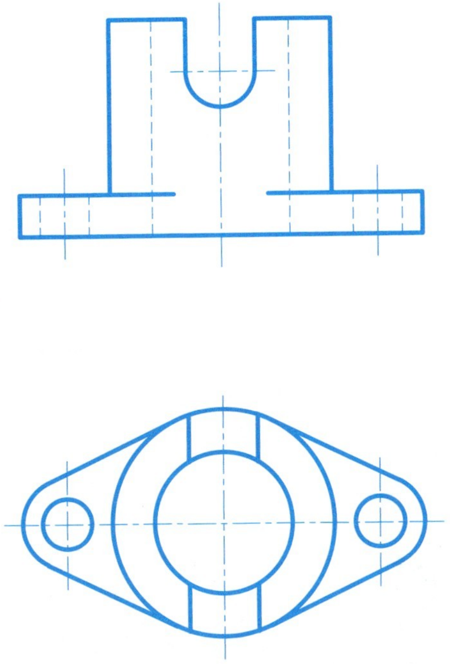

5-6到5-7
参考答案

5-6到5-7
动画／模型

| Chapter 5 General Principles of Representation 第5章 图样画法 | Name | No. |

5-6 Draw a half sectional front view at the set position.　在指定位置作半剖的主视图。

5-7 Draw a half sectional front view and a full sectional left-side view at the set positions. 在指定位置作半剖主视图和全剖左视图。

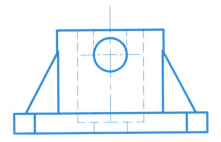

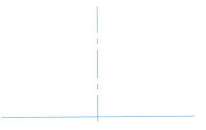

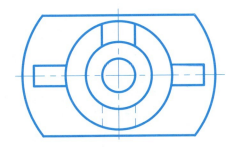

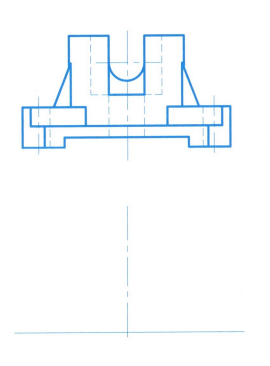

5-8到5-9
参考答案

5-8到5-9
动画／模型

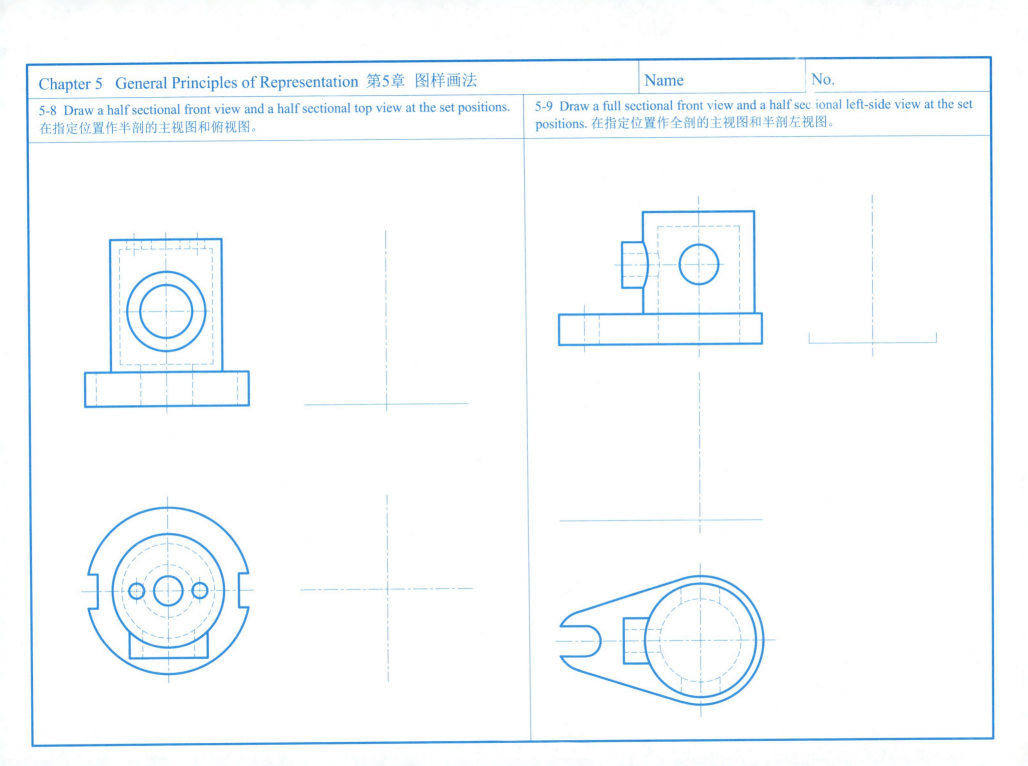

5-10到5-11
参考答案

5-10到5-11
动画／模型

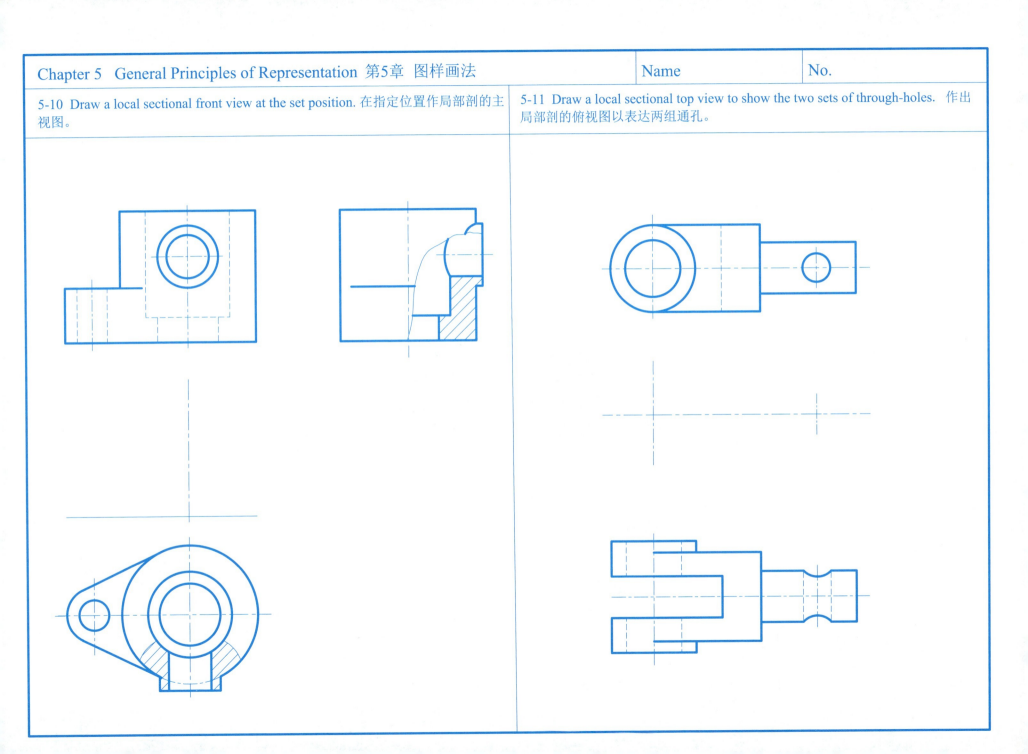

5-12到5-13
参考答案

5-12到5-13
动画／模型

| Chapter 5 General Principles of Representation 第5章 图样画法 | Name | No. |

5-12 Draw a full sectional front view with three paralleling cutting planes. 画出用三平行平面剖切的全剖主视图。

5-13 Draw a full sectional front view with two intersecting planes. 画出用两相交平面剖切的全剖主视图。

5-14(1-2)
参考答案

5-14(1-2)
动画／模型

| Chapter 5 General Principles of Representation 第5章 图样画法 | Name | No. |

5-14 Draw a full sectional front view wtih two paralleling cutting planes. 画出用两平行平面剖切的全剖主视图。

(1)

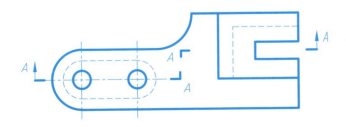

(2)

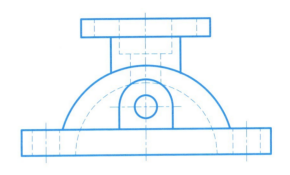

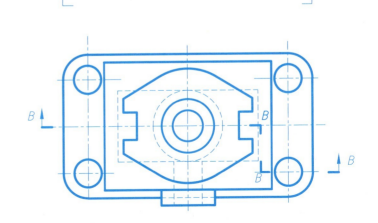

5-15到5-16
参考答案

5-15到5-16
动画／模型

| Chapter 5 General Principles of Representation 第5章 图样画法 | Name | No. |

5-15 Draw a superposition cut view at the set position. 在指定位置画出重合断面图。

5-16 Draw a removed cut view at the set position. 在指定位置画出移出断面图。

5-17到5-18
参考答案

5-17到5-18
动画／模型

| Chapter 5　General Principles of Representation　第5章　图样画法 | Name | No. |

5-17　Draw cut views (on the scale of 1∶1) and partial enlargement views (on the scale of 2∶1) at the set positions.　在指定位置作出断面图（比例为1∶1）和局部放大图(比例为2∶1)。

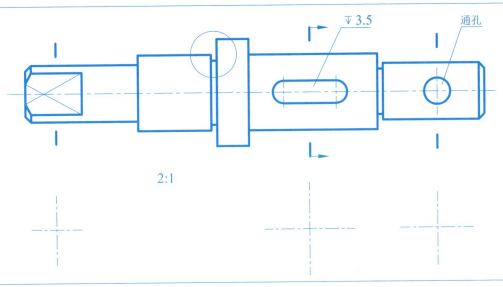

2:1

5-18　Add the missing details. 补画剖视图中缺少的图线。

(1)

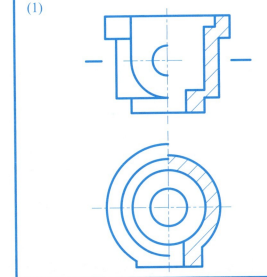

(2)

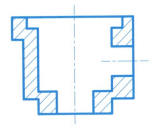
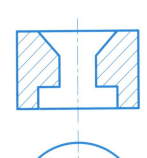
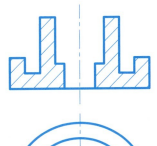

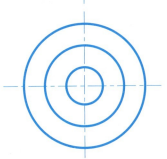
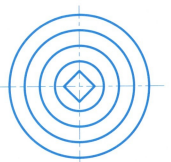

5-19(1-2) 参考答案   5-19(1-2) 动画／模型

| Chapter 5   General Principles of Representation 第5章 图样画法 | Name | No. |

5-19 Use appropriate representation methods to describe the components clearly on appointed locations and dimension the components (take sizes from the drawings and hold integer values). 选用适当的表达方法在指定位置将机件表达清楚，并标注尺寸（尺寸数值从图中量取整数）。

(1)

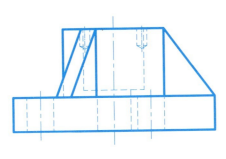

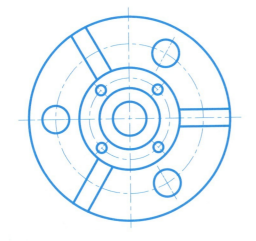

(2)

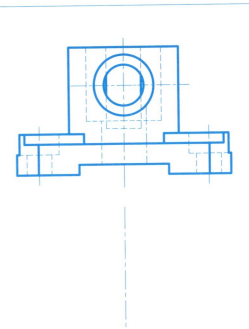

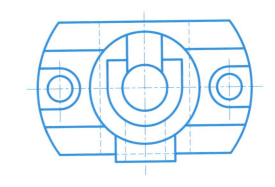

6-1
参考答案

| Chapter 6 Standard Parts and Commonly Used Parts 第6章 标准件与常用件 | Name | No. |

6-1 Draw the screw threads and mark. 画出螺纹并标注螺纹的规定代号。

(1) Trapezoidal screw thread, d=18 mm, P_h=8 mm, n=2, left-hand. 梯形螺纹，大径为18mm，导程为8 mm，线数为2，左旋。

(2) Coarse plain thread, d=18 mm, P=2.5 mm, the thread length is 30 mm. 粗牙普通螺纹，大径为18 mm，螺距为2.5 mm，螺纹长度为30 mm。

(3) Fine plain thread, D=12 mm, P=1 mm, the thread length is 26 mm. 细牙普通螺纹，大径为12 mm，螺距为1 mm，螺纹长度为26 mm。

(4) Pipe thread with 55 degree thread angle where pressure-tight joints are not made on the thread, the nominal diameter is 1/2 inch, the tolerance grade is B, the thread length is 30 mm. 非密封的管螺纹，公称直径为1/2 in，B级，右旋，螺纹长度为30 mm。

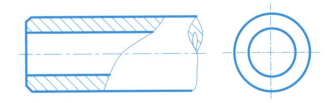

6-2
参考答案

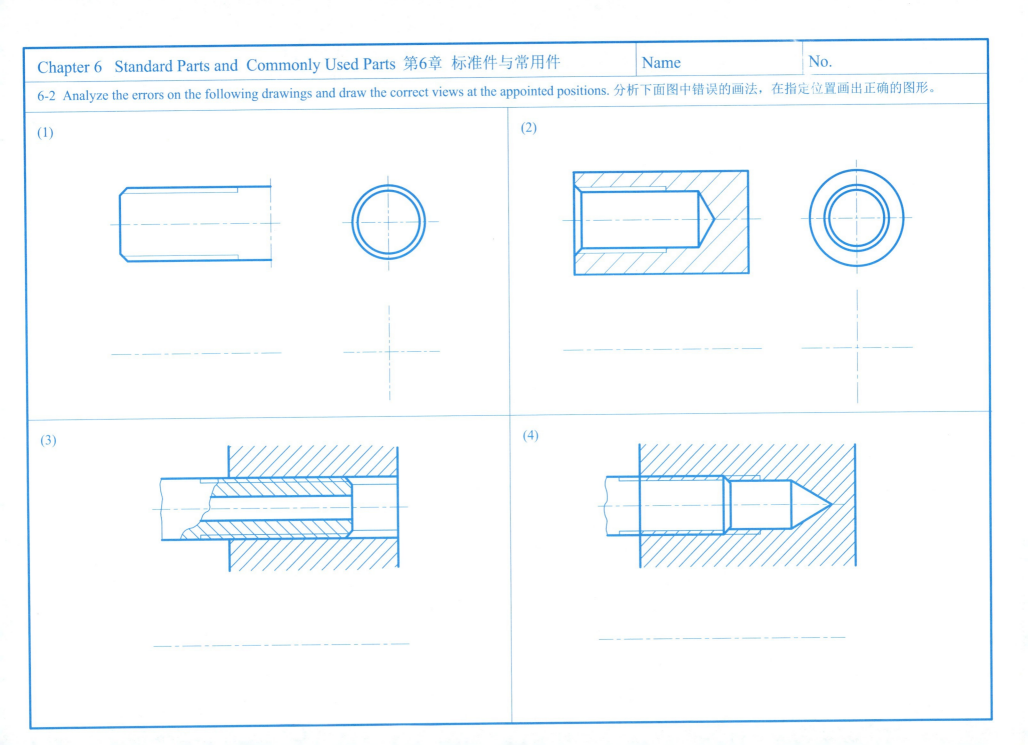

6-3
参考答案

| Chapter 6 Standard Parts and Commonly Used Parts 第6章 标准件与常用件 | Name | No. |

6-3 Draw a section view of the internal and external thread joint and the cut view $A-A$, given that the length of thread engagement is 15 mm. 画内、外螺纹的连接图，旋合长度为15 mm, 并作$A-A$断面图。

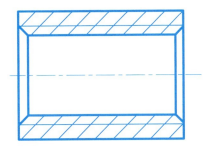

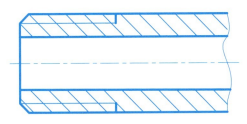

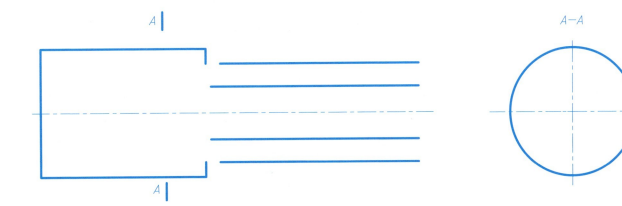

6-4
参考答案

| Chapter 6 Standard Parts and Commonly Used Parts 第6章 标准件与常用件 | Name | No. |

6-4 Complete the marking of each standard part and dimension the parts. 写出各标准件标记并标注尺寸。

(1) Hexagon head bolt, product grade C, d=12 mm, l=45 mm. 六角头螺栓，C级，螺纹规格为d=12 mm, 公称长度为45 mm。

(2) Double end stud, d=16 mm, l=50 mm, l_1=20 mm, b_m= 1.25 d. 双头螺柱，螺纹规格(两端)为d=16 mm，公称长度l=50 mm，旋入长度l_1=20 mm，b_m=1.25 d。

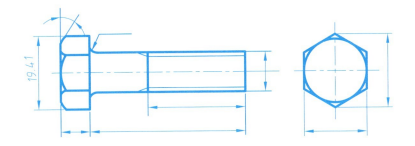

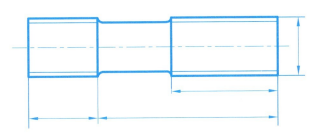

(3) Hexagon nut, style 1, product grade A, D=16 mm. 1型六角螺母，A级，螺纹规格为d=16 mm。

(4) Slotted countersunk flat head screw, d=8 mm, l=40 mm. 开槽沉头螺钉，螺纹规格为d=8 mm, 公称长度为40 mm。

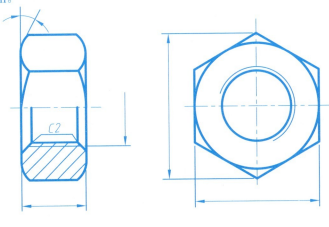

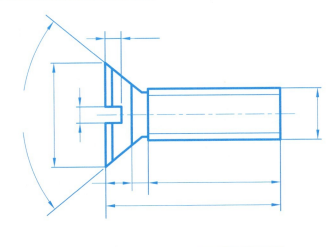

6-5到6-6
参考答案

| Chapter 6 Standard Parts and Commonly Used Parts 第6章 标准件与常用件 | Name | No. |

6-5 Draw the three views of the bolted joints according to the conventional representation. 按照规定画法作出螺栓连接的三视图。

6-6 Draw the two views of the screw joint on an appropriate scale given that the assembly is medium. 以合适的比例作出螺钉连接的主视图和俯视图。（中等装配精度）

Specified symbols of the parts
各零件的规定标记：
螺柱　　GB/T 5782 M16×55
螺母　　GB/T 6170 M16
垫圈　　GB/T 97.2 16

6-7到6-8
参考答案

| Chapter 6 Standard Parts and Commonly Used Parts 第6章 标准件与常用件 | Name | No. |

6-7 Complete the cylindrical gear drawing and dimension ($m=2$ mm, $Z=24$). 补全圆柱齿轮的图形并标注尺寸 ($m=2$ mm, $Z=24$)。

6-8 Complete the cylindrical gear mating drawing and dimension ($m=2$ mm, $Z_1=32$, $Z_2=21$). 补全圆柱齿轮啮合图 ($m=2$ mm, $Z_1=32$, $Z_2=21$)。

6-9
参考答案

| Chapter 6   Standard Parts and Commonly Used Parts 第6章 标准件与常用件 | Name | No. |

6-9 Complete the following views and mark dimensions of the keyway given that the lenghth of the square and rectangular key of type A is 35 mm .The diameter of the shaft hole should be measured from the drawing and the drawing scale is 1:2.The other sizes should be determined by reference to the tables. 已知齿轮和轴用A型普通平键连接，补全下列各视图并标注键槽尺寸。轴孔直径从图上量取（比例为1:2），键的长度为36 mm。其他尺寸需查表确定。

Specified symbol of the key 键的规定标记：_____

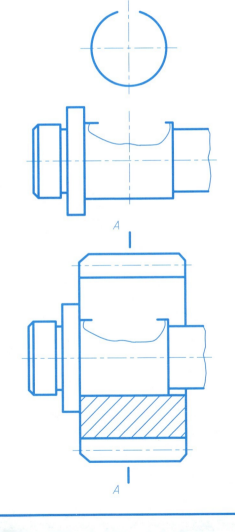

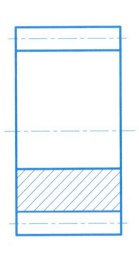

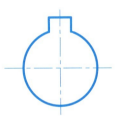

A—A

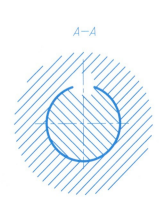

6-10
参考答案

| Chapter 6 Standard Parts and Commonly Used Parts 第6章 标准件与常用件 | Name | No. |

6-10 Complete the following views given that the length of the pin is 40 mm. 已知齿轮和轴用B型圆柱销连接，销的长度为40 mm。

(1) Select the appropriate cylindrical pin and give the specified mark of the pin. 选择合适圆柱销并写出销的规定标记。
   Specified mark of the pin 键的规定标记_____

(2) Complete the full sectional view of the pin connection with appropriate proportions, the sizes should be determined by reference to the relevant tabel. 查表确定销的尺寸，用适当比例补全销连接的全剖视图。

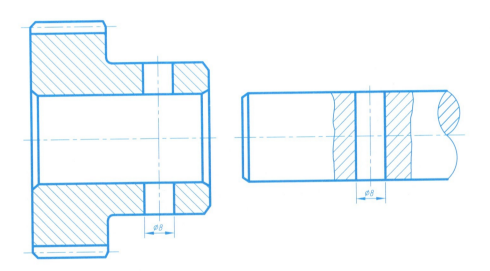
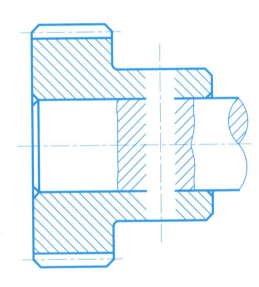

6-11到6-13
参考答案

| Chapter 6 Standard Parts and Commonly Used Parts 第6章 标准件与常用件 | Name | No. |

6-11 Given that the diameters of the supporting shoulders at both ends of the shaft are 25 mm and 15 mm respectively, draw the two rolling bearings at the set positions with the simplified drawing method and the bearings should be drawn on a suitable scale. 已知轴两端支承肩处的直径分别为25 mm和15 mm，用规定画法，以合适比例画出支承处的滚动轴承。

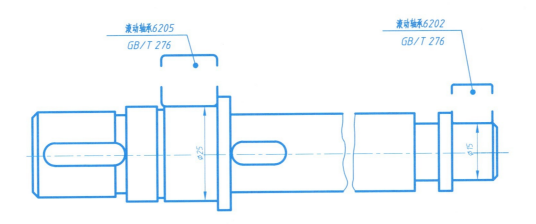

滚动轴承6205 GB/T 276   滚动轴承6202 GB/T 276

6-12 Draw the view of the key and dimension according to the specified symbol. 根据规定标记，画出键的视图并标注尺寸。	6-13 Draw the view of the pin and dimension according to the specified symbol. 根据规定标记，画出销的视图并标注尺寸。
GB/T 1096 键 12×8×40	GB/T 119.1 销 12m6×40

7-1
参考答案

Chapter 7 Detail Drawings 第7章 零件图

7-1 Label the surface roughness symbols of the appointed surfaces in the views according to the given *Ra* values. 根据给定的*Ra*值，将指定表面粗糙度代号标注在视图上。

(1)

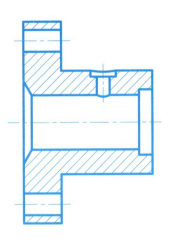

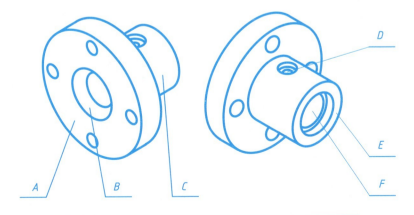

A	*B*	*C*	*D*	*E*	*F*	其余
3.2 μm	1.6 μm	12.5 μm	6.3 μm	12.5 μm	25 μm	

(2)

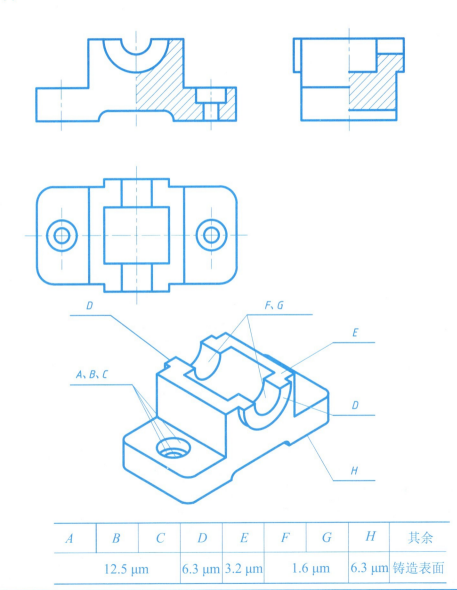

A	*B*	*C*	*D*	*E*	*F*	*G*	*H*	其余
12.5 μm			6.3 μm	3.2 μm	1.6 μm		6.3 μm	铸造表面

7-2
参考答案

Chapter 7 Detail Drawings 第7章 零件图

7-2 Label the indicated surface roughness symbols of the appointed surfaces in the view according to the given Ra values. 根据给定的Ra值，在视图上标注各指定表面的粗糙度。

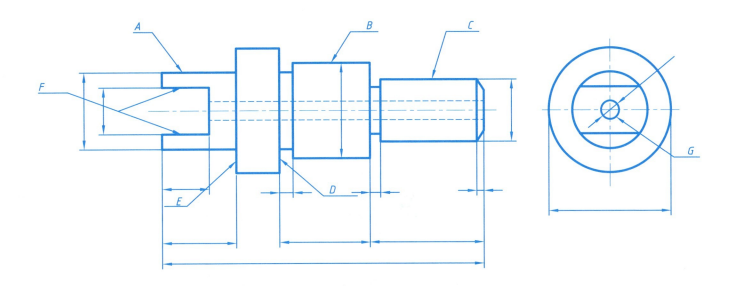

A	B	C	D	E	F	G	其余
3.2 μm	0.4 μm	1.6 μm	12.5 μm		6.3 μm		3.2 μm

7-3到7-4
参考答案

| Chapter 7 Detail Drawings 第7章 零件图 | Name | No. |

7-3 Label the indicated surface roughness symbols according to the given *Ra* values.
根据给定的*Ra*值，在视图上标注表面粗糙度。

(1) The surface roughness(*Ra*) of the φ15 inner hole is 1.6 μm. φ15内孔*Ra*值为1.6 μm。
(2) The surface roughness(*Ra*) of the countersunk holes is 12.5 μm. 沉孔*Ra*值为12.5 μm。
(3) The surface roughness(*Ra*) of the base plane and the two end faces with a spacing of 16 mm is 6.3 μm. 间距16 mm的两端面以及底平面*Ra*值均为6.3 μm。
(4) The remaining casting surfaces do not need cuting, and can maintain the original casting supply condition. 其余铸造表面不需要切削加工，保持原铸件供应状况。

7-4 Label the indicated surface roughness symbols according to the given *Ra* values.
根据给定的*Ra*值，在视图上标注各表面粗糙度。

(1) The surface roughness(*Ra*) of the outer surface of the φ30 and φ28 cylinders are 1.6 μm. φ30、φ28圆柱外表面*Ra*值为1.6 μm。
(2) The surface roughness(*Ra*) of the working surface of the M20 screw thread is 12.5 μm. 螺纹工作表面*Ra*值为12.5 μm。
(3) The surface roughness(*Ra*) of the keyway work surfaces are 3.2 μm, the surface roughness(*Ra*) of the keyway undersurface is 6.3 μm. 键槽工作表面*Ra*值为3.2 μm，底面*Ra*值为6.3 μm。
(4) The surface roughness(*Ra*) of the φ4 taper pin hole is 1.6 μm. φ4锥销孔表面*Ra*值为1.6 μm。
(5) The surface roughness(*Ra*) of the remaining surfaces after cutting is 12.5 μm. 其余表面切削加工后，*Ra*值为12.5 μm。

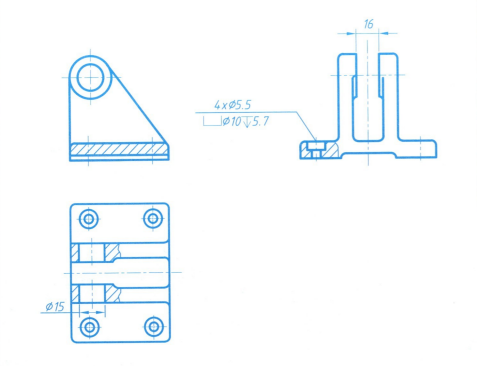

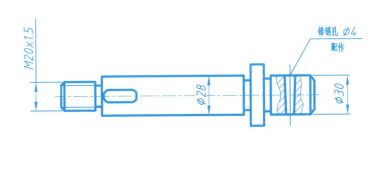

7-5(1-4)
参考答案

| Chapter 7 Detail Drawings 第7章 零件图 | Name | No. |

7-5 Mark the symbols for the geometric tolerances. 标注下面几何公差的符号。

(1) The generatrix straightness tolerance of the cylinder is 0.1 mm. 圆柱素线的直线度公差为0.1 mm。

(2) The flatness tolerance of the upper surface of the part is 0.2 mm. 机件上表面的平面度公差为0.2 mm。

(3) The roundness tolerance of the normal section of the round table is 0.1 mm. 圆台正截面的圆度公差为0.1 mm。

(4) The small hole is parallel to the large hole (the reference) with a tolerance of $\phi 0.03$ mm for parallelism. 小孔平行于大孔（基准），其平行度公差为$\phi 0.03$ mm。

7-5(5-8)
参考答案

| Chapter 7　Detail Drawings　第7章　零件图 | Name | No. |

(5) The vertical surface is perpendicular to the bottom surface of the object (the reference) with a tolerance of 0.1 mm for perpendicularity. 竖直表面垂直于物体的底面(基准)，其垂直度公差为0.1 mm。

(6) The ture position tolerance of the two holes is $\phi 0.8$ mm. 两孔的位置度公差为$\phi 0.8$ mm。

50

(7) The small cylinder is coaxial with the large cylinder (the reference), and the coaxiality tolerance is $\phi 0.1$ mm. 小圆柱与大圆柱(基准)同轴，其同轴度公差为$\phi 0.1$ mm。

(8) The notch is symmetrical to the central plane (the reference) of the upper and lower surfaces with a symmetry tolerance of 0.1 mm. 切口对称于上、下表面的中心平面(基准)，其对称度公差为0.1 mm。

7-6
参考答案

Chapter 7 Detail Drawings 第7章 零件图

7-6 Mark the size deviation values of the fitting surfaces of the parts (The size deviation values should be determined by reference to the relevant tables). 查表标注出零件配合面的尺寸偏差值。

(1) $\phi 20H7/g6$, $\phi 32H7/k6$

(2) $\phi 47JS7$, $\phi 20k6$, $\phi 47JS7/f8$

7-7
参考答案

Chapter 7 Detail Drawings 第7章 零件图

7-7 Mark the fitting codes on the assembly drawing according to the nominal sizes and the upper and lower deviation values of the fitting parts given in the part drawing.
已知零件图配合部位的公称尺寸与上、下偏差值，在装配图上标注配合代号。

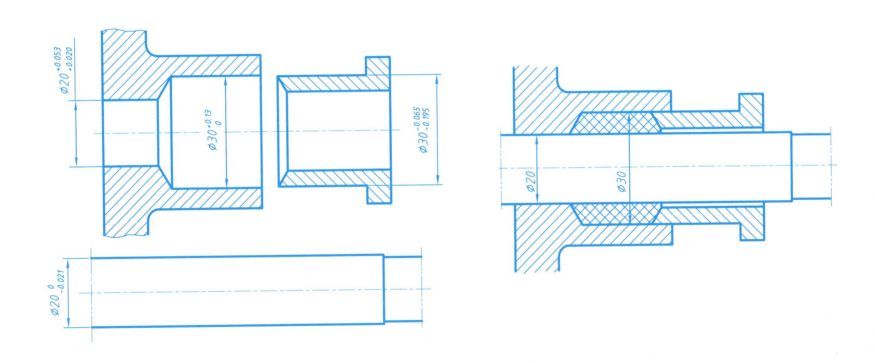

7-8
参考答案

Chapter 7 Detail Drawings 第7章 零件图

7-8 According to the nominal sizes and fit codes marked on the assembly drawing, indicate the fit systems, the types of fits and the standard tolerance grades of the holes and shafts, and mark the size deviation values on the part drawing. 根据装配图中所注的公称尺寸和配合代号指出图中的配合采用的基准制及配合类型，轴、孔的标准公差等级，并在零件图上标注尺寸偏差值。

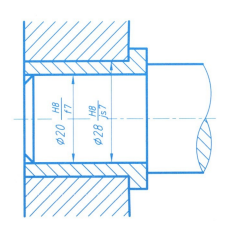

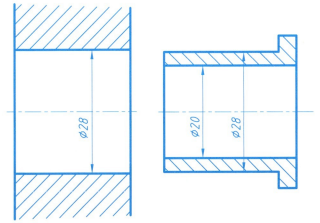

Ø20H8/f7:

Ø28H8/js7:

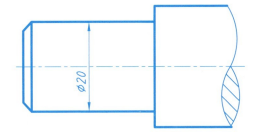

7-9
动画／模型

Chapter 7 Detail Drawings 第7章 零件图

Name	No.

7-9 Draw the detail drawing of the fork(see page 68). 绘制拨叉零件图(见P68)。

1)Content 内容

According to the stereogram, select an appropriate proportion to draw the part drawing on an A3 sized sheet. 根据立体图，选择合适的比例，在A3图纸上绘制零件图。

2)Purpose 目的

①Further improve the ability of flexible use of the expression methods of machine parts through practice. 通过练习进一步提高灵活运用机件表达法的能力。

②Master the methods and steps of constructing detail drawings. 掌握绘制零件图的方法与步骤。

3)Requirements 要求

①The part should be expressed completely, clearly and reasonably. 零件图表达完整、清楚、合理。

②Size marking of the part shall meet the drawing standards and requirements, and must be complete, clear, correct and reasonable. 零件的尺寸标注符合制图标准与要求，做到标注完整、清晰、正确与合理。

③Technical requirements only need to be marked as given on the stereogram. 技术要求只需标注立体图上所给出的即可。

④The drawing quality of lines, arrows, various symbols and others is good. 图线、尺寸箭头及各种符号等绘制质量较好。

4)Method guidance 方法指导

①Understand the stereogram and structure of the part, and ignore small structures when analyzing the shape. 看懂立体图以及零件的结构，做形体分析时可忽略小的结构。

②Select appropriate expression methods according to the physical structure, such as view, section, etc. 根据形体结构选择适当的表达方法，如视图、剖视图等。

③Common process structures, such as keyway, have fixed expression methods. 常见的工艺结构(如键槽等)有固定的表达方法。

5)Marking of surface roughness Ra value 表面粗糙度Ra值的标注

①The Ra values of surfaces A, B, C and I are 12.5 μm, the Ra values of surfaces D、E and F are 3.2 μm, the Ra value of surface G is 1.6 μm and the Ra value of surface H is 6.3 μm. 面A、面B、面C、面I的Ra值为12.5 μm，面D、面E和面F为3.2 μm，面G的Ra值为1.6 μm，面H的Ra值为6.3 μm。

②The Ra value of the keyway of the hub is 6.3 μm on the side and 12.5 μm on the top. 轮毂的键槽侧面的Ra值为6.3 μm，顶面的Ra值为12.5 μm。

③The rest is cast surface. 其余面为铸造表面。

6)Other 其他

The part material is HT200. 零件材料为HT200。

7-9
参考答案

| Chapter 7 Detail Drawings 第7章 零件图 | Name | No. |

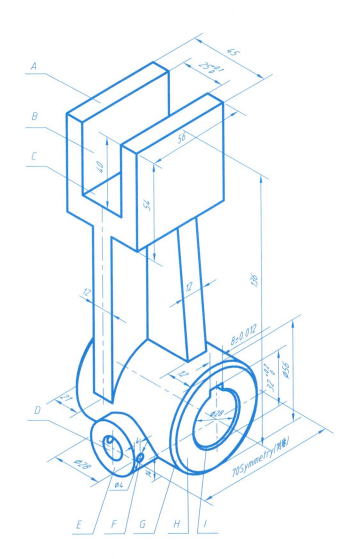

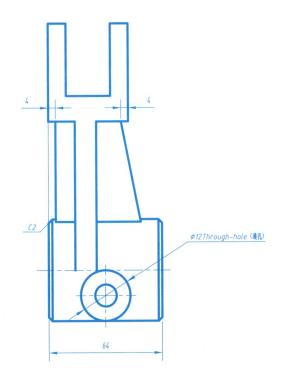

TECHNICAL REQT 技术要求
All unspecified chamfers must be C1. 未注倒角C1。

7-10
参考答案(1)

| Chapter 7   Detail Drawings   第7章 零件图 | Name | No. |

7-10 Study the detail drawing of the stand-off (see page 70) carefully, then finish the exercises. 看懂托脚零件图(见P70)并完成习题。

(1) Draw the left view. 画出左视图。

(2) Note the roughness symbols of surfaces *a* and *b* on the left view. 在左视图上注出表面*a*、*b*的粗糙度代号。

(3) ⊥ ⌀0.05 A indicates that the measured feature is _____, the geometric feature is _____, the inspection item is _____, the tolerance value is _____ . ⊥ ⌀0.05 A 表示被测要素是_____，基准要素是_____，几何特征是_____，公差值是_____。

(4) ⌀26H9 indicates that the nominal size is _____, the tolerance class is _____, the tolerance grade is _____, the fundamental deviation code is _____, and the hole _____ (is/isn't) benchmark holes. ⌀26H9表示公称尺寸是_____，公差带代号是_____，公差等级是_____，基本偏差代号是_____，该孔_____(是/不是)基准孔。

(5) The M6-6H thread is _____ thread, 6H refers to _____, and the screw direction is_____ _____. M6-6H螺纹是_____螺纹，6H指_____，螺纹旋向为_____。

(6) Note the main dimension datums in each direction in the drawing. 在图中注出各个方向上的主要尺寸基准。

(7) ⌀26H9 can be expressed as $\phi 26^{+0.052}_{0}$ in terms of the limit deviation value. Then the maximum value of the bore diameter is_____, the minimum value is _____, the upper deviation is _____, the tolerance is _____. 采用极限偏差数值来标注，可将⌀26H9表示为$\phi 26^{+0.052}_{0}$，则孔径的最大值为_____，最小值为_____，上偏差为_____，公差为_____。

7-10
参考答案(2)

7-10
动画／模型

| Chapter 7 Detail Drawings 第7章 零件图 | Name | No. |

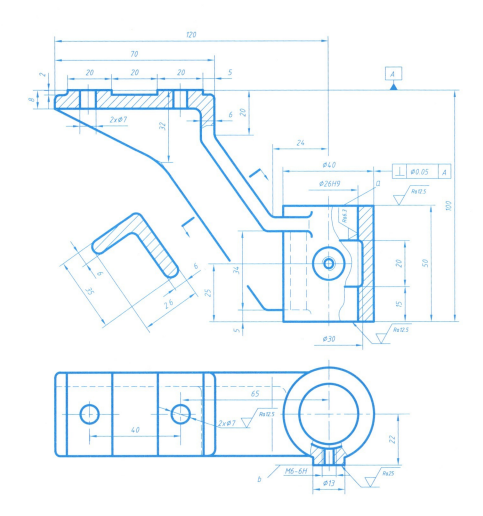

TECHNICAL REQT 技术要求
1. All unspecified fillet radii must be *R*2~*R*3. 未注圆角R2~R3。
2. The casting shall not have pores, cracks and other defects. 铸件不得有气孔、裂纹等缺陷。

Designed 设计		Stand-off 托脚	Scale 比例	
Checked 校核			quantity 数量	
			HT150	

7-11
参考答案(1)

Chapter 7 Detail Drawings 第7章 零件图	Name	No.

7-11 Study the detail drawing of the turret mount(see page 72) carefully, then finish the exercises. 看懂回转架零件图并完成习题。

(1) Draw the half-section view $B-B$ at the specified position. 在指定的位置画出$B-B$半剖视图。

(2) Point out the dimension dutams in the length, width and height directions in the drawing. 在图中指出长、宽、高三个方向的尺寸基准。

(3) There are _____ kinds of surface roughness requirements for this part, surfaces with the highest and lowest surface quality requirements are _____ and _____ respectively. 该零件的表面粗糙度要求有 _____ 种，表面质量要求最高和最低的面分别是 _____ 和 _____ 。

(4) Is the keyway size in the hole of Φ22H9 up to standard? _____. The width and depth of the keyway in the shaft that mates with the hole should be (by reference to relevant standards) _____ and _____ respectively. Φ22H9孔内的键槽尺寸是否符合标准？ _____ 。与该孔相配合的轴的键槽尺寸的宽和深应该分别为（查相关标准） _____ 和 _____ 。

(5) The nominal size of $41^{+0.05}_{0}$ is _____, and the upper limit of size is _____, the lower limit of size is _____. $41^{+0.05}_{0}$表示公称尺寸为 _____ ，上极限尺寸为 _____ ，下极限尺寸为 _____ 。

(6) Try to explain the meaning of ∜(√). A: _____. 试解释 ∜(√) 的含义。

答：_____。

7-11
参考答案(2)

7-11
动画／模型

Chapter 7 Detail Drawings 第7章 零件图

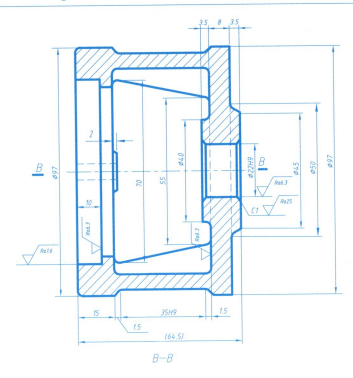

B–B

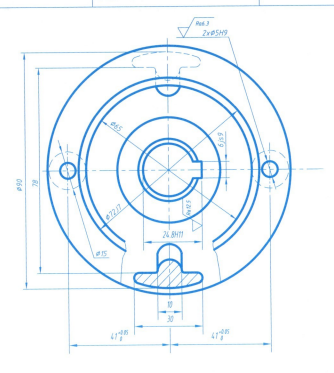

TECHNICAL REQT 技术要求
1. All unspecified fillet radii must be R2~R3. 未注圆角R2~R3。
2. The casting shall not have pores, cracks and other defects. 铸件不得有气孔、裂纹等缺陷。

Turret mount 回转架

HT200

7-12
参考答案(1)

Chapter 7　Detail Drawings　第7章　零件图	Name	No.

7-12　Study the detail drawing of the hollow shaft (see page 74) carefully, then finish the exercises. 看懂空心轴零件图并完成习题(见P74)。

(1) Draw the cut view on $B-B$ at the specified position. 在指定的位置画出 $B-B$ 断面图.

(2) This part belongs to _____, and the front view conforms to the _____ position of the part. 该零件为 _____ 类零件，主视图符合零件的 _____ 位置。

(3) The front view is _____ view. Besides, there are _____ view and _____ view in the drawing. 主视图是 _____ 视图；此外，该零件图中还有 _____ 图和 _____ 图。

(4) The left end face of the part has _____ holes, the _____ is 8 mm, the _____ depth is 10 mm, the _____ depth is 12 mm. 零件左端面有 ___ 个 ____ 孔， _____ 为8 mm、 _____ 深为10 mm、 _____ 深为12 mm。

(5) The surface roughness of the $\phi 95h6$ cylinder is obtained using the method of _____ material, the upper limit of Ra of the cylinder is _____; the diameter of the hole marked ① is _____, the upper limit of Ra of the hole surface is _____. $\phi 95h6$ 圆柱面的粗糙度是用 _____ 材料的方法获得，该圆柱面的 Ra 上限值为 _____；图中标记①的孔的直径为 _____，该孔的表面粗糙度 Ra 上限值为 _____。

(6) There is a distance of _____ between the two dashed lines at the position marked ② in the view. 在图中标记②的部位，两条虚线的距离为 _____。

(7) The curve marked ③ in the figure is the intersection line formed by the intersection of _____ and _____. 图中标记③的曲线是 _____ 和 _____ 相交而形成的 _____。

(8) The wire frame marked ④ in the figure has two size dimensions of _____ and a location dimension of _____. 图中标记④的线框，其定形尺寸是 _____，定位尺寸是 _____。

(9) The $\phi 132\pm 0.2$ outer cylindrical surface can be processed to the maximum diameter of _____, the minimum diameter of _____, and the tolerance is _____. $\phi 132\pm 0.2$ 外圆柱面可加工到直径最大为 _____，最小为 _____，公差为 _____。

7-12
参考答案(2)

7-12
动画／模型

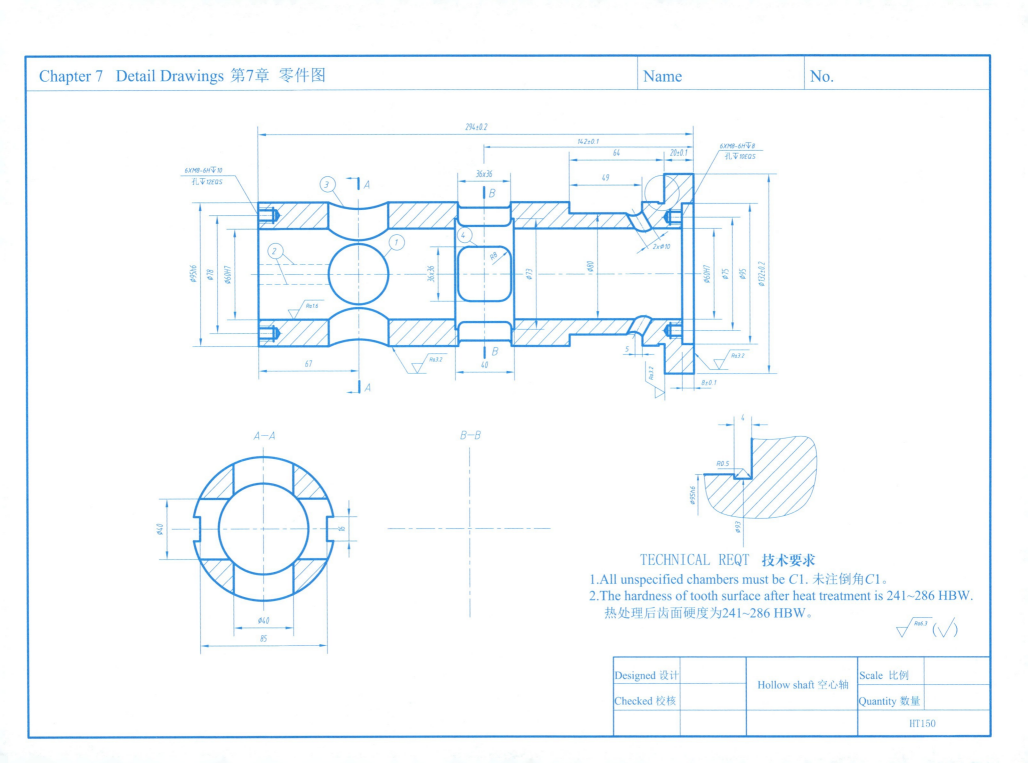

7-13
参考答案(1)

| Chapter 7 Detail Drawings 第7章 零件图 | Name | No. |

7-13 Study the base drawing(see page 76) carefully, then finish the exercises. 看懂底座零件图并完成习题（见P76）。

(1)This part belongs to _____, the front view conforms to the _____ position of the part. 该零件为_____类零件，主视图符合零件的_____位置。

(2)This detail drawing adopts the following expression methods: _____ view, _____ view, _____ view and _____ view. 该零件图采用的表达方法有：_____视图、_____视图、_____视图和_____视图。

(3)Draw the left view of the part at the specified position. 在指定位置画出零件的左视图。

(4)There are _____ kinds of surface roughness requirements for this part, which are _____. 该零件有_____种表面粗糙度要求，它们分别是

_____。

(5)Point out the dimension datums in the directions of length, width and height in the drawing. 在图中指出长、宽、高三个方向的尺寸基准。

7-13
参考答案(2)

7-13
动画 / 模型

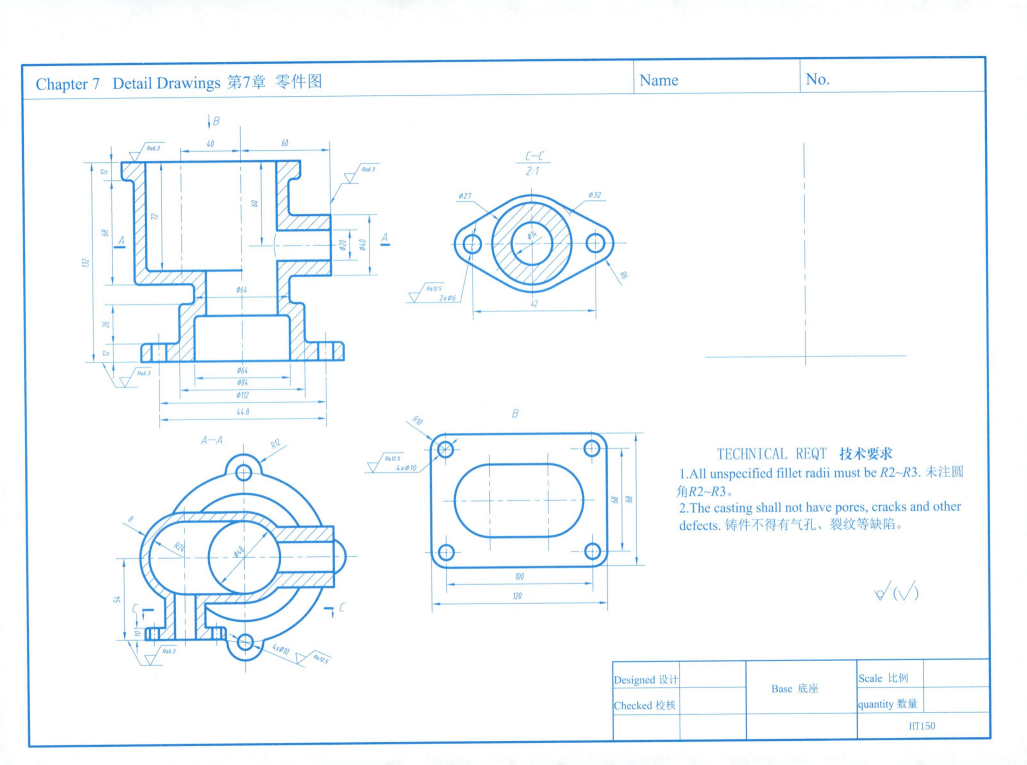

8-1
参考答案

Chapter 8　Assembly Drawings　第8章　装配图	Name	No.

8-1 Draw an assembly drawing of the gear pump. 绘制齿轮泵装配图。

1)Content 内容

Draw an assembly drawing on an A2 sized sheet according to the part drawings(see page 79-83) of the gear pump . 根据所给的齿轮泵各零件的零件图(见P79~P83),在A2图纸上拼绘装配图。

2)Purpose 目的

Be familiar with the contents of assembly drawing and the general provisions of assembly drawing expression, master the methods and steps of drawing assembly drawings, and dimensioning of assembly drawings. 熟悉装配图的内容及装配图表达的一般规定，掌握绘制装配图的方法、步骤，以及装配图的尺寸标注。

3)Requirements 要求

①The projection are correct and the expression scheme is concise. 投影正确，表达方案简洁。

②The assembly relation of parts is expressed clearly and correctly. 零件装配关系表达清楚，正确。

③Dimensioning, item list, part numbers and other marks meet the requirements. 尺寸标注、明细栏、零件编号及其他标注符合要求。

4)Method guidance 方法指导

①Before drawing, we should understand all the part drawings, understand the working principle of the assembly and the assembly relationships between the parts. 绘制前应看懂所有的零件图，了解装配体的工作原理和各零件之间的装配关系。

②Select the appropriate expression scheme according to the characteristics of the assembly. In addition to common expressions way of the parts, consider special expression methods of assembly drawing and strive to express clearly and succinctly. 根据装配体的特点选择合适的表达方案，除了常用的机件表达方式外，考虑采用装配图的特殊表达法，力求表达清晰，简洁。

③Only dimensions such as specification dimensions, performance dimensions, assembly dimensions, and installation dimensions need to be marked when dimensioning. 标注尺寸时只需要标注规格、性能、装配和安装等几类尺寸。

④Note the drawing of thread connections and section lines. For the same part, no matter where it appears, the direction and spacing of its section line are the same. 注意螺纹连接的画法和剖面线的画法。对于同一个零件，不论它在哪里出现，其剖面线方向与间隔均相同。

8-1(1)
动画／模型

8-1(2)
动画／模型

Chapter 8　Assembly Drawings　第8章　装配图

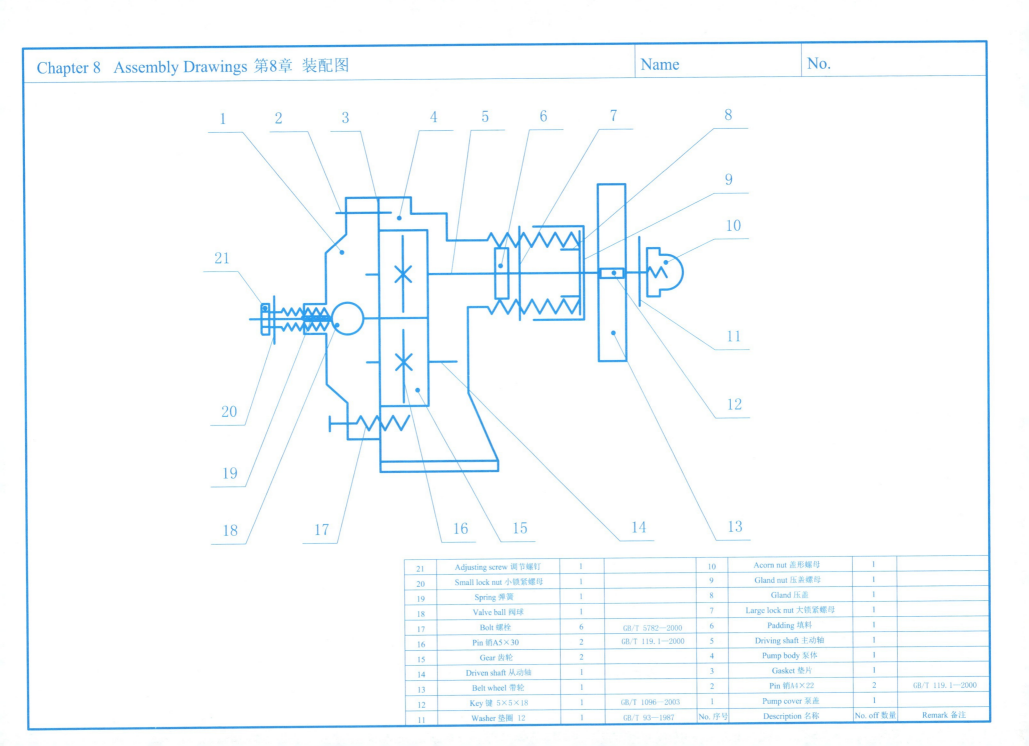

21	Adjusting screw 调节螺钉	1		10	Acorn nut 盖形螺母	1	
20	Small lock nut 小锁紧螺母	1		9	Gland nut 压盖螺母	1	
19	Spring 弹簧	1		8	Gland 压盖	1	
18	Valve ball 阀球	1		7	Large lock nut 大锁紧螺母	1	
17	Bolt 螺栓	6	GB/T 5782—2000	6	Padding 填料	1	
16	Pin 销A5×30	2	GB/T 119.1—2000	5	Driving shaft 主动轴	1	
15	Gear 齿轮	2		4	Pump body 泵体	1	
14	Driven shaft 从动轴	1		3	Gasket 垫片	1	
13	Belt wheel 带轮	1		2	Pin 销A4×22	2	GB/T 119.1—2000
12	Key 键 5×5×18	1	GB/T 1096—2003	1	Pump cover 泵盖	1	
11	Washer 垫圈 12	1	GB/T 93—1987	No. 序号	Description 名称	No. off 数量	Remark 备注

8-1(3)
动画／模型

Chapter 8 Assembly Drawings 第8章 装配图

All unspecified fillet radii must be R3. 未注圆角R3。

No. 序号	Description 名称	Scale 比例	Material 材料	No. off 数量
1	Pump cover 泵盖	1:1	HT200	1

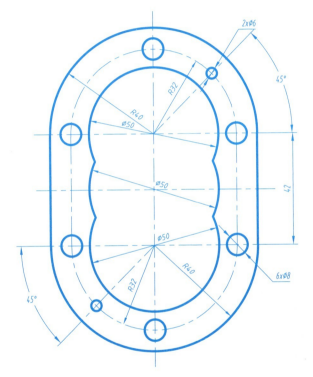

The thickness is 0.5 mm. 厚度为0.5 mm。

No. 序号	Description 名称	Scale 比例	Material 材料	No. off 数量
3	Gasket 垫片	1:1	Asbestos 石棉	1

8-1(4)
动画／模型

Chapter 8   Assembly Drawings   第8章 装配图

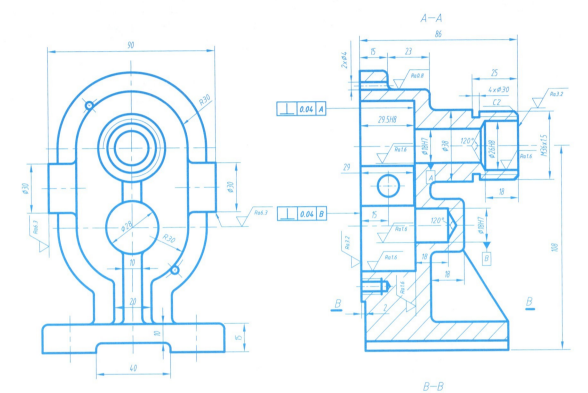

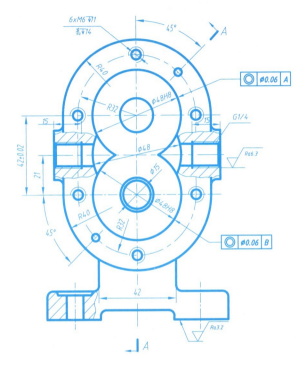

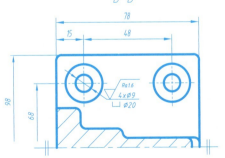

TECHNICAL REQT   技术要求

1. All unspecified fillet radii must be $R3 \sim R5$. 未注圆角 $R3 \sim R5$。
2. The surface roughness of the chamfer and the retreat groove is $Ra12.5$. 倒角、退刀槽的表面粗糙度均为 $Ra12.5$。
3. The parallelism tolerance of the two $\phi 18H7$ hole axes is not greater than 0.05 mm. 两个 $\phi 18H7$ 孔轴线的平行度公差不大于 0.05 mm。

No. 序号	Description 名称	Scale 比例	Material 材料	No. off 数量
4	Pump body 泵体	1:1	HT200	1

8-1(5)
动画 / 模型

Chapter 8 Assembly Drawings 第8章 装配图

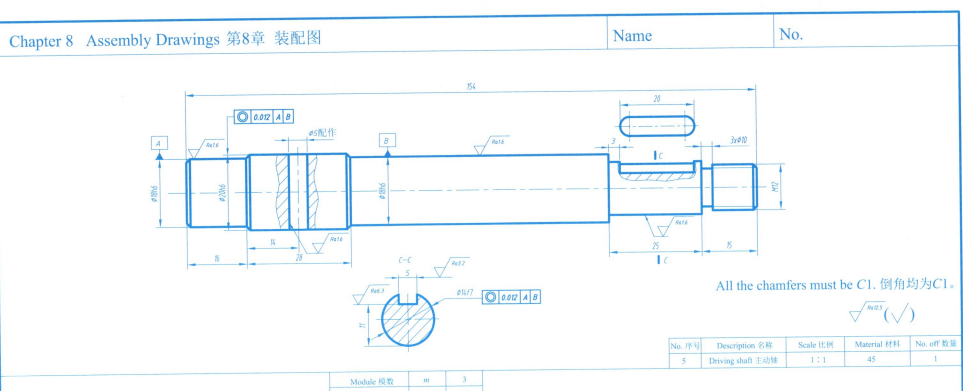

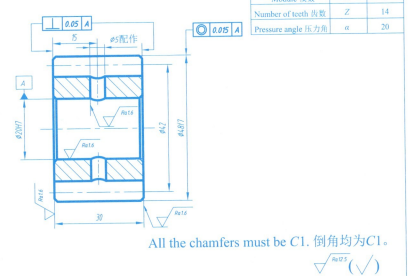

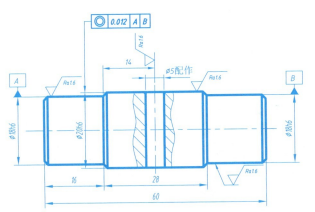

8-1(6)
动画／模型

Chapter 8   Assembly Drawings 第8章 装配图

Name	No.

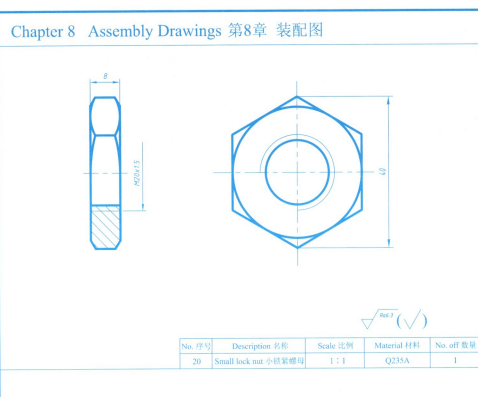

No. 序号	Description 名称	Scale 比例	Material 材料	No. off 数量
20	Small lock nut 小锁紧螺母	1:1	Q235A	1

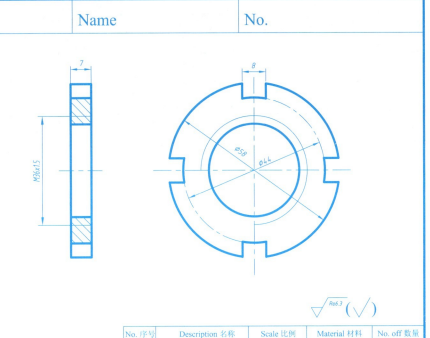

No. 序号	Description 名称	Scale 比例	Material 材料	No. off 数量
7	Large lock nut 大锁紧螺母	1:1	Q235A	1

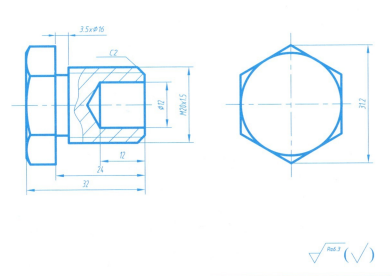

No. 序号	Description 名称	Scale 比例	Material 材料	No. off 数量
21	Adjusting screw 调节螺钉	1:1	Q235A	1

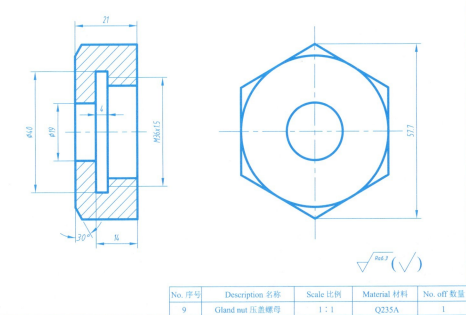

No. 序号	Description 名称	Scale 比例	Material 材料	No. off 数量
9	Gland nut 压盖螺母	1:1	Q235A	1

8-1(7)
动画／模型

Chapter 8   Assembly Drawings 第8章 装配图

Name	No.

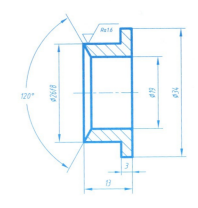

No. 序号	Description 名称	Scale 比例	Material 材料	No. off 数量
8	Gland 压盖	1:1	35	1

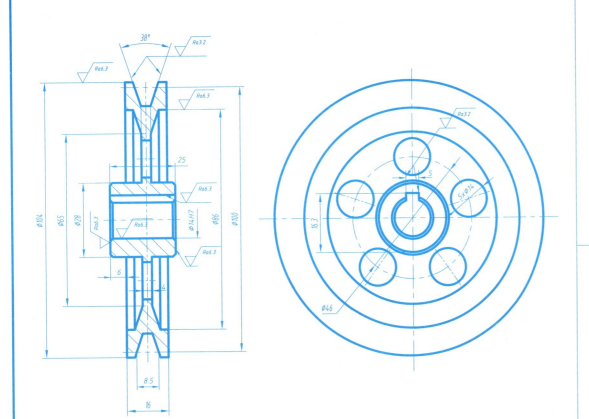

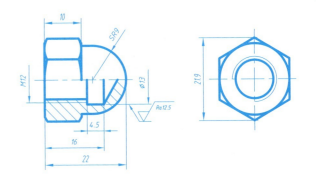

All unspecified chamfers must be C1. 未注倒角C1。

No. 序号	Description 名称	Scale 比例	Material 材料	No. off 数量
13	Belt wheel 带轮	1:1	ZL-3	1

No. 序号	Description 名称	Scale 比例	Material 材料	No. off 数量
10	Acorn nut 盖形螺母	1:1	Q235A	1

8-2(1)
动画／模型

8-2(1)
参考答案

Chapter 8　Assembly Drawings　第8章　装配图

| | | Name | No. |

8-2　Draw an assembly drawing of the flat nose clamp according to the detail drawings(see page 85-87). 由平口钳零件图(见P85~P87)绘制装配图。

Working principle: Flat nose clamp is used to fix pieces for machining. It mainly contains a fixed clamp body, a movable clamp body, a clamp plank, a screw rod and a sheathing nut. Turn the screw rod fitted on the fixed clamp body, the sheathing nut, connected with the movable clamp body by screws, will move straight. Therefore, turning the screw rod can drive the movable clamp body along the fixed clamp body, which will make the mouth of the clamp open or close to clamp or unload the part.

工作原理：平口钳是一种用于夹持工件的部件，它主要由固定钳身、活动钳身、钳口板、丝杠和套螺母等组成。丝杠固定在固定钳身上，转动丝杠可带动套螺母做直线运动，套螺母与活动钳身用螺钉连成整体。当丝杠转动时，活动钳身就会沿着固定钳身移动，实现钳口闭合或开放，从而夹紧或松开工件。

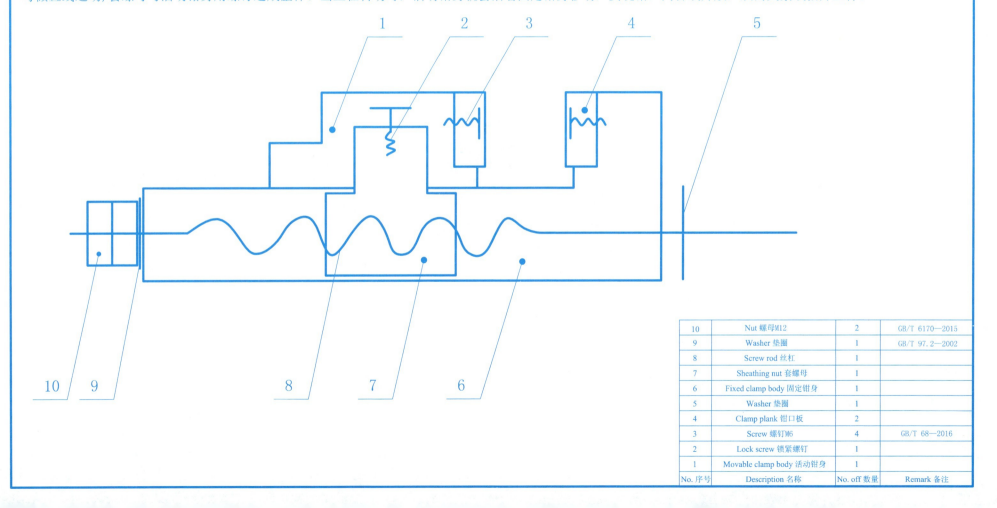

No. 序号	Description 名称	No. off 数量	Remark 备注
10	Nut 螺母M12	2	GB/T 6170—2015
9	Washer 垫圈	1	GB/T 97.2—2002
8	Screw rod 丝杠	1	
7	Sheathing nut 套螺母	1	
6	Fixed clamp body 固定钳身	1	
5	Washer 垫圈	1	
4	Clamp plank 钳口板	2	
3	Screw 螺钉M6	4	GB/T 68—2016
2	Lock screw 锁紧螺钉	1	
1	Movable clamp body 活动钳身	1	

8-2(2)
动画／模型

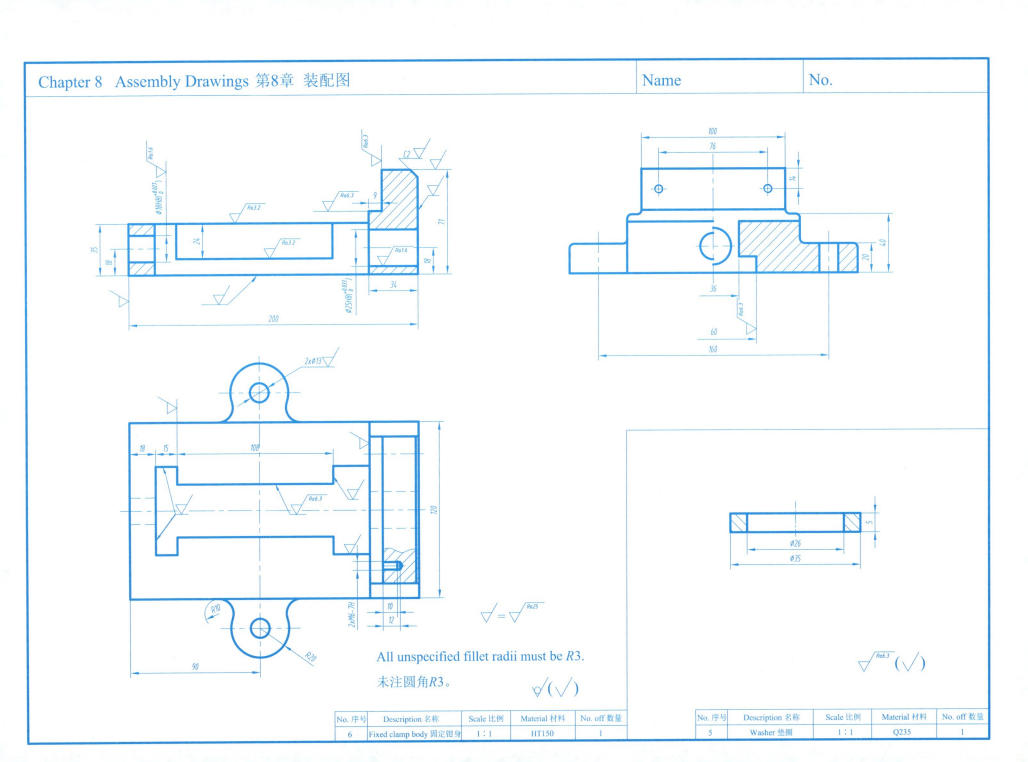

8-2(3)
动画 / 模型

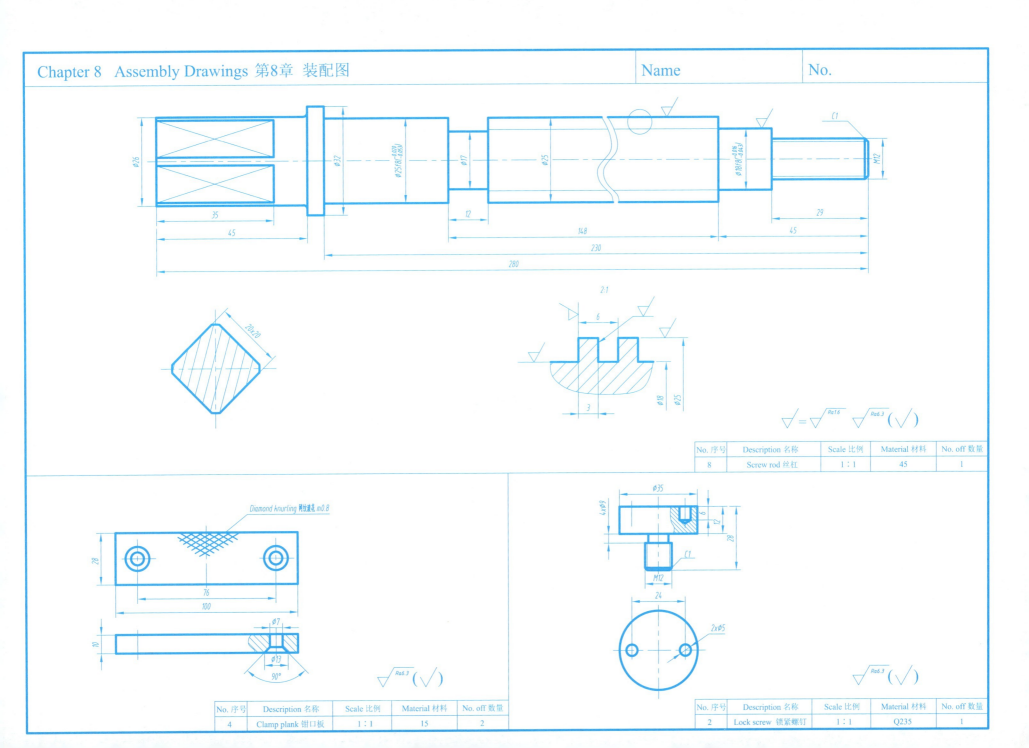

8-2(4)
动画／模型

Chapter 8 Assembly Drawings 第8章 装配图

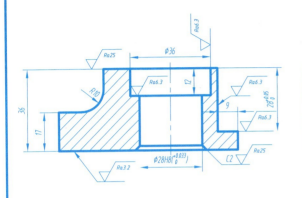
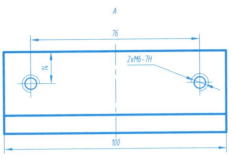
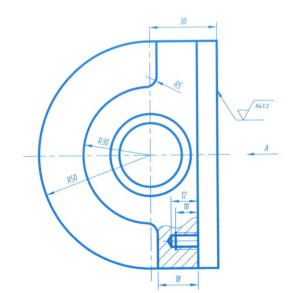

All unspecified fillet radii must be R3. 未注圆角R3。

No. 序号	Description 名称	Scale 比例	Material 材料	No. off 数量
1	Movable clamp body 活动钳身	1:1	HT150	1

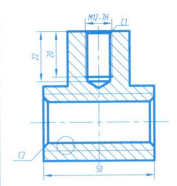
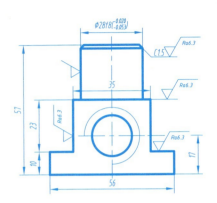
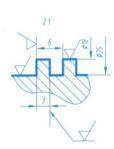

No. 序号	Description 名称	Scale 比例	Material 材料	No. off 数量
7	Sheathing nut 套螺母	1:1	HT150	1

8-3(1)
动画／模型

8-3(1)
参考答案

Chapter 8　Assembly Drawings　第8章　装配图	Name	No.

8-3 Remove and draw the safety valve parts. 拆画安全阀零件图。

1) Content 内容

　　Read the safety valve assembly drawing(see page 89) and draw the detail drawing of part 1. The required dimensions are to be taken directly from the drawing. 读安全阀装配图(见P89)并画出零件1的零件图，所需尺寸直接从图中量取。

2) Principle Specification 原理说明

　　The unit is installed in pipelines to control the flow of liquid for safety protection. 该部件安装在管道上用于控制液体流量，从而起到安全保护作用。

　　When working, the liquid enters the valve body 1 from the $\phi 20$ pipeline on the right side of the safety valve, and flows out from the $\phi 20$ pipeline on the lower side of the safety valve. If the hydraulic pressure of the liquid transmitted in the pipeline rises, then valve 3 rises, and the liquid will flow out of the $\phi 20$ hole on the left side of the safety valve to guarante the safe operation of the equipment. 工作时液体从安全阀右侧的$\phi 20$管道中进入阀体1，从安全阀下侧的$\phi 20$管道中流出。如果管道中所传送的液体压力升高，则阀门3将上升，液体将从安全阀左侧的$\phi 20$孔中流出，从而保证设备安全运行。

　　The working pressure in valve body 1 can be adjusted by adjusting screw 9 to change the degree of compaction of spring 2 .Bonnet 10 plays a protective role. 通过调节螺杆9，改变弹簧2的压紧程度，可调节阀体1中的工作压力。阀盖10起防护作用。

8-3(2)
动画／模型

Chapter 8   Assembly Drawings   第8章 装配图

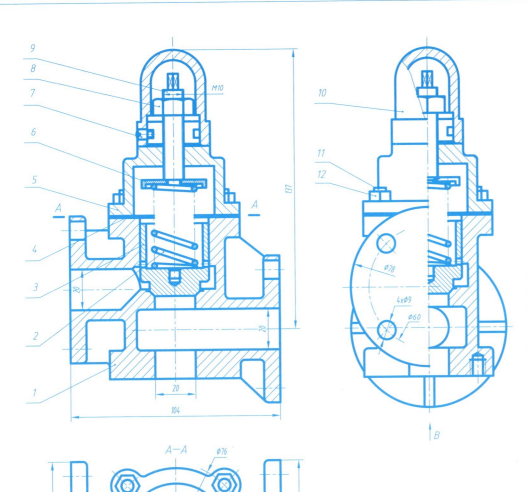

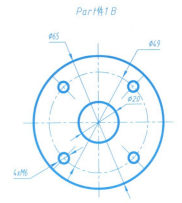

Part件1B

TECHNICAL REQT 技术要求
1. The joint surface between the value and the value body needs to be ground to prevent water leakage and air leakage. 阀门与阀体间的结合面需研磨，不漏水和气。
2. Green paint should be applied to raw surfaces. 未加工表面涂绿色油漆。

12	Nut 螺母M6	4	GB/T 6170—2015
11	Double end stud 双头螺柱M6×20	4	GB 898—1988
10	Bonnet 阀帽	1	
9	Screw 螺杆	1	
8	Nut 螺母M10	1	GB/T 6172.1—2016
7	Bolt 螺钉M6×6	1	GB 75—85
6	Tray 托盘	1	
5	Valve 阀盖	1	
4	Gasket 垫片	1	
3	Valve 阀门	1	
2	Spring 弹簧	1	GB/T 65—2000
1	Valve body 阀体	1	
No. 序号	Description 名称	No. off 数量	Remark 备注

Designed 设计		Date 日期		(Matl mark 材料标记)	(School name 校名)
Checked 校核					
Approved 审核				Scale 比例 1:1	Safety valve 安全阀
Class name 班级		STU No. 学号		Total 共 张 No. 第 张	(DRG No. 图号)

Chapter 9　Computer Drawing　第9章　计算机绘图

Name　　　No.

Complete the following drawings and dimension with the Auto CAD software. 使用AutoCAD软件绘制下列图形并标注尺寸。

(1)

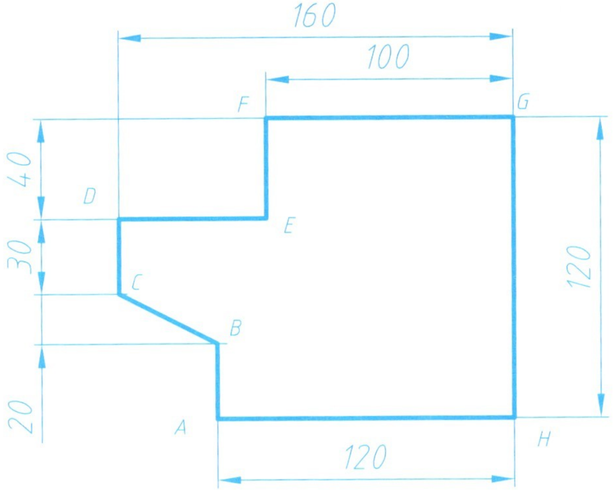

(2)

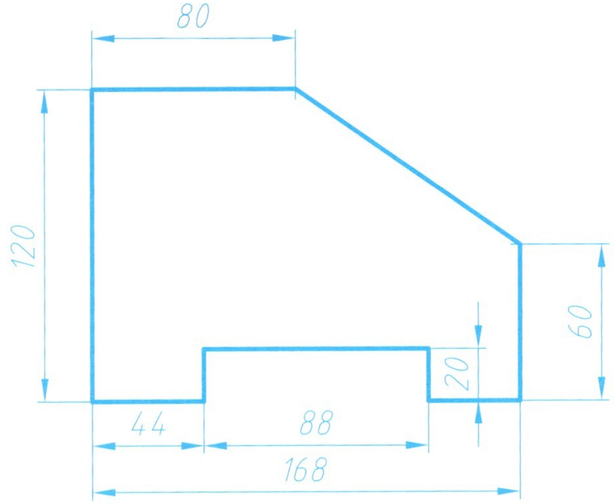

(3)

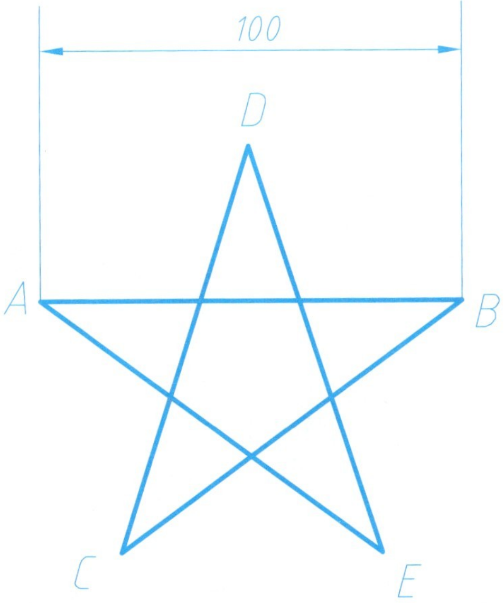

(4)

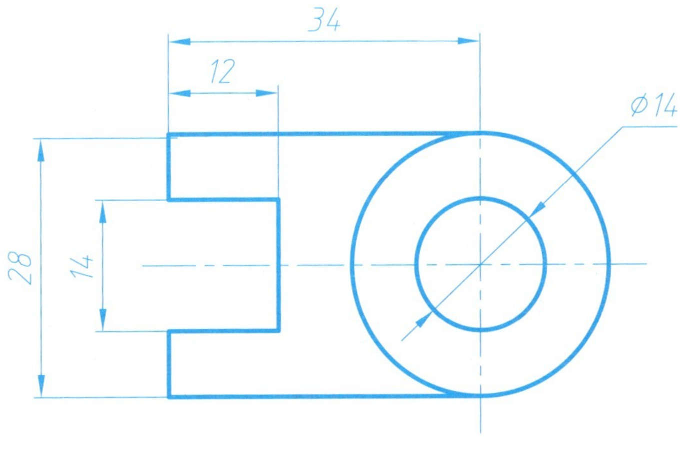

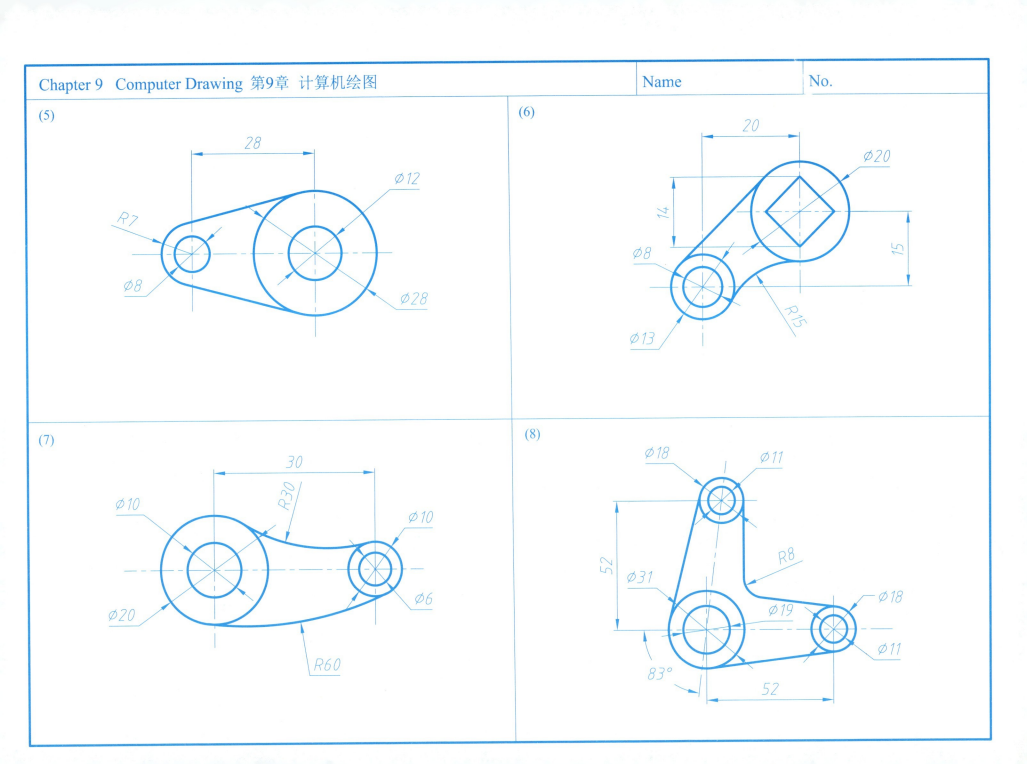

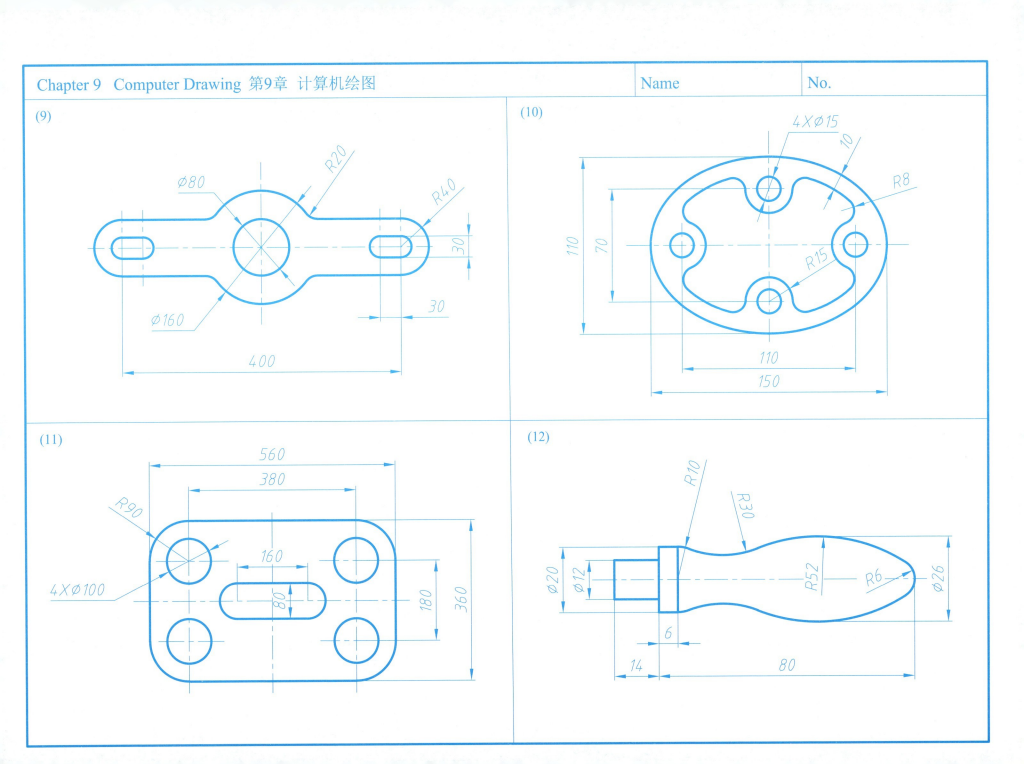

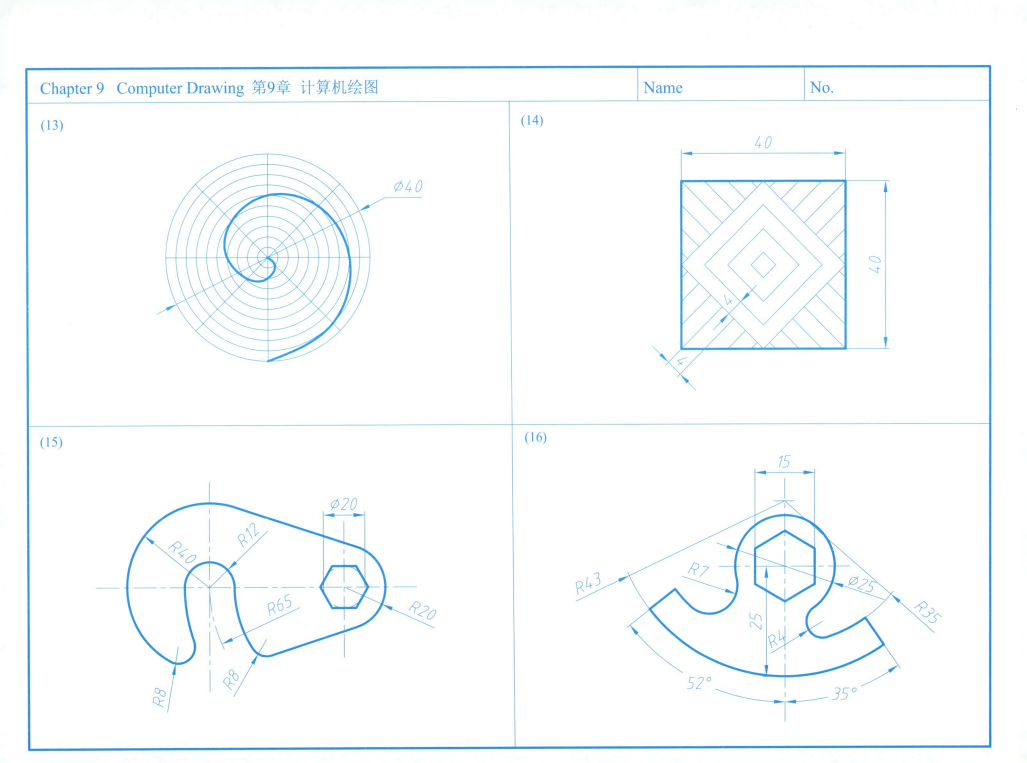

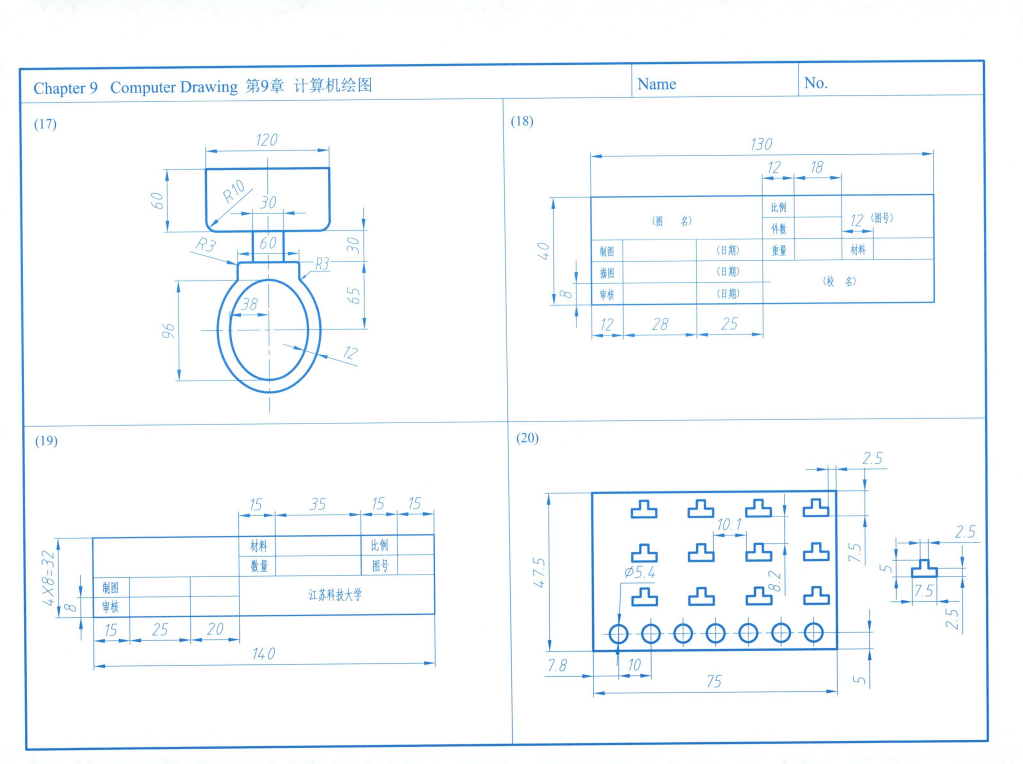

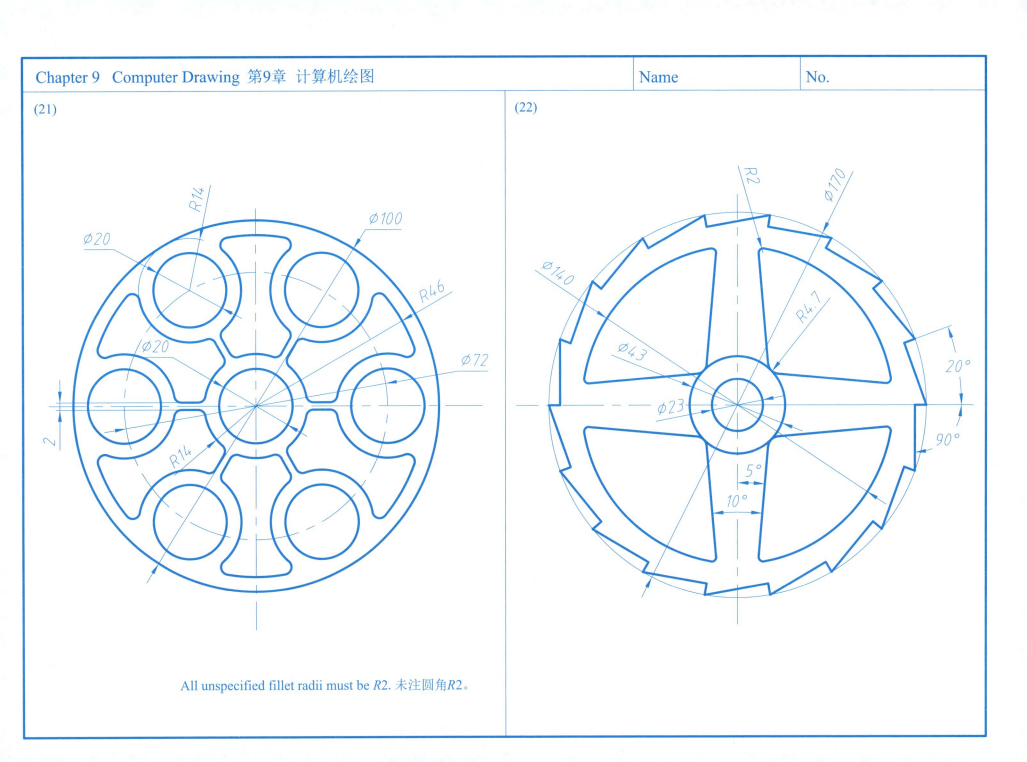

| Chapter 9 Computer Drawing 第9章 计算机绘图 | Name | No. |

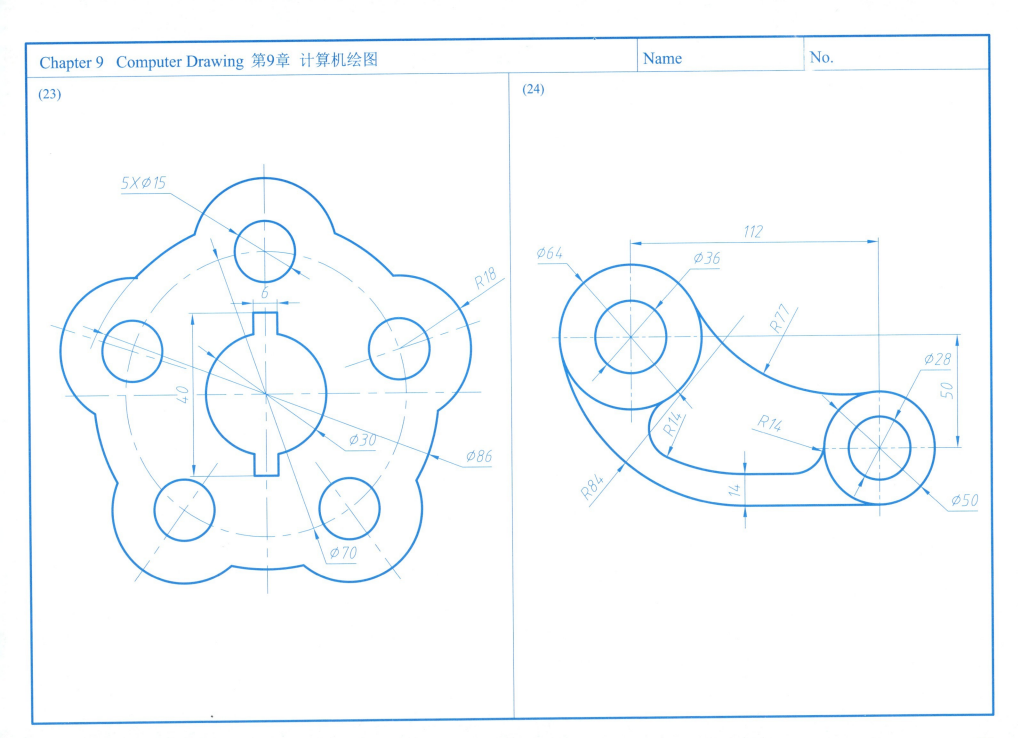

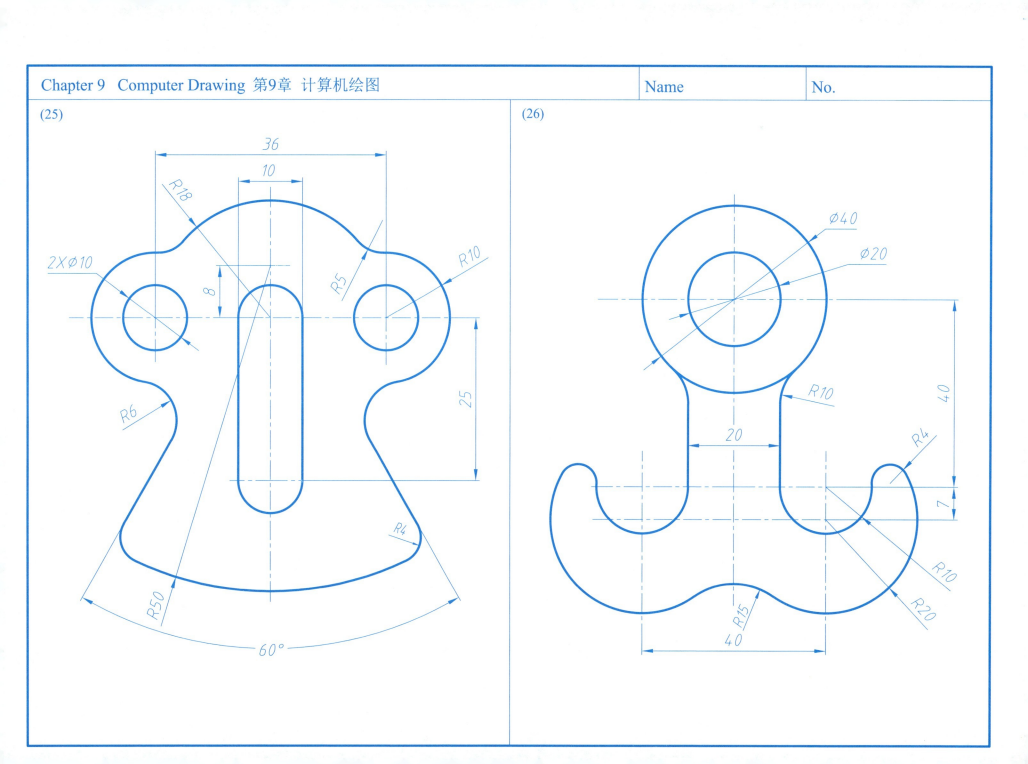

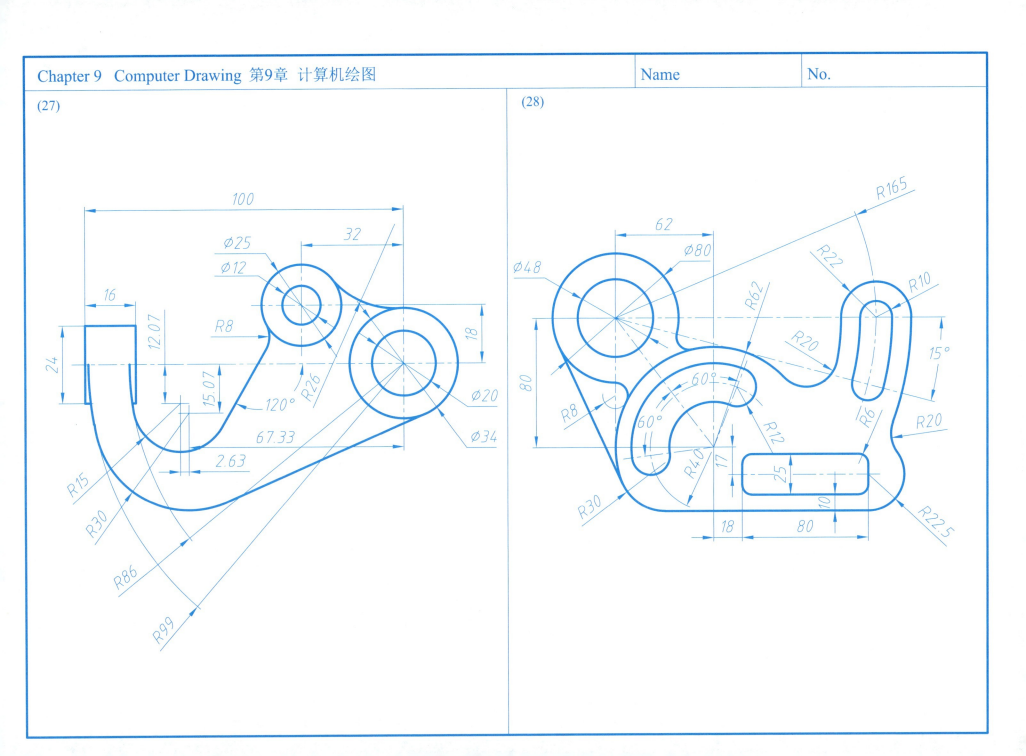

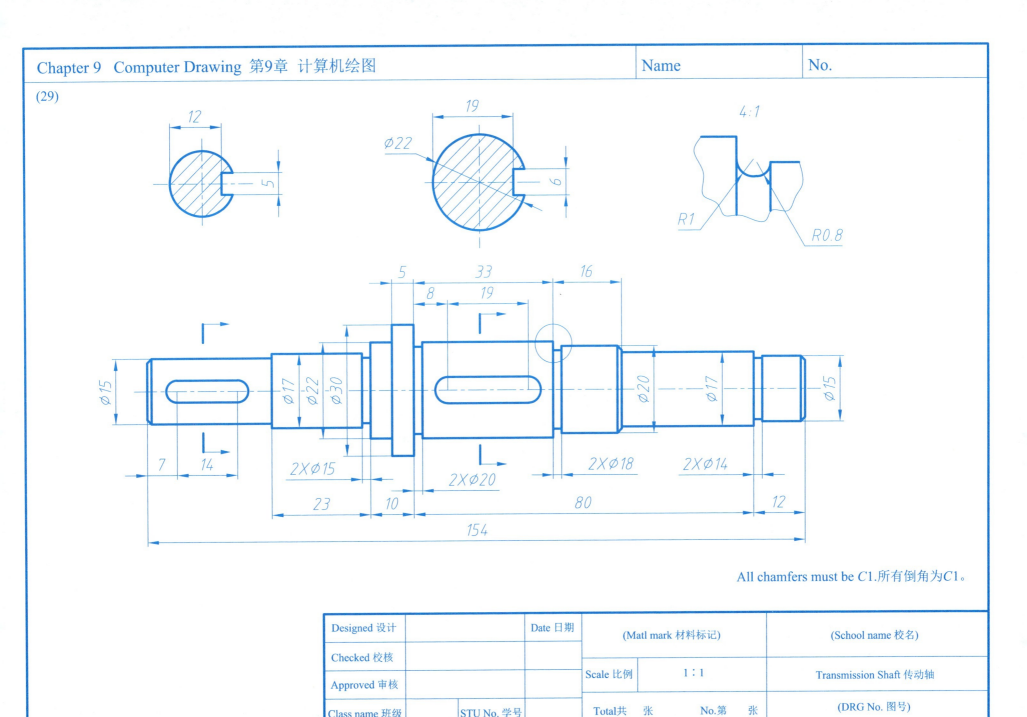

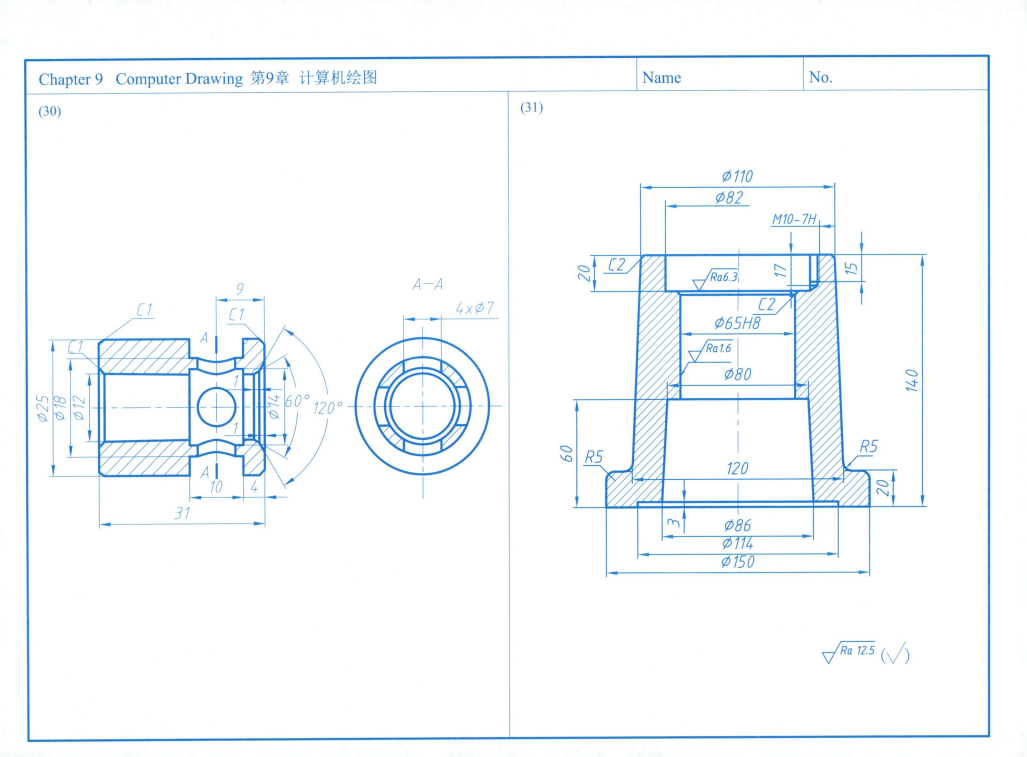

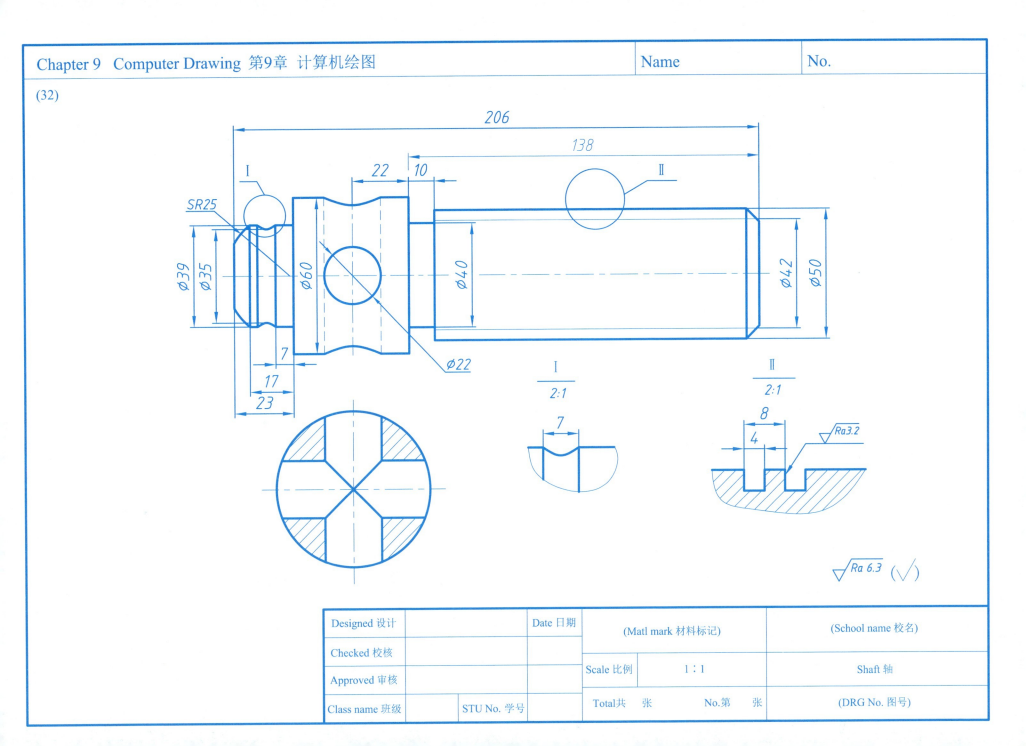

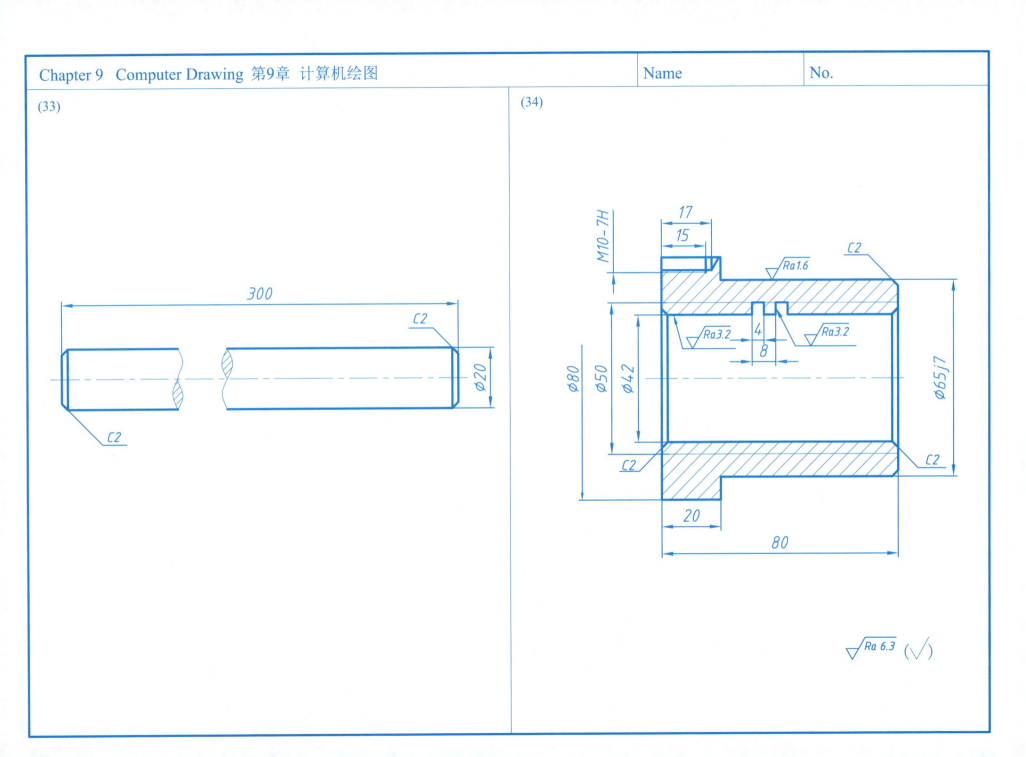

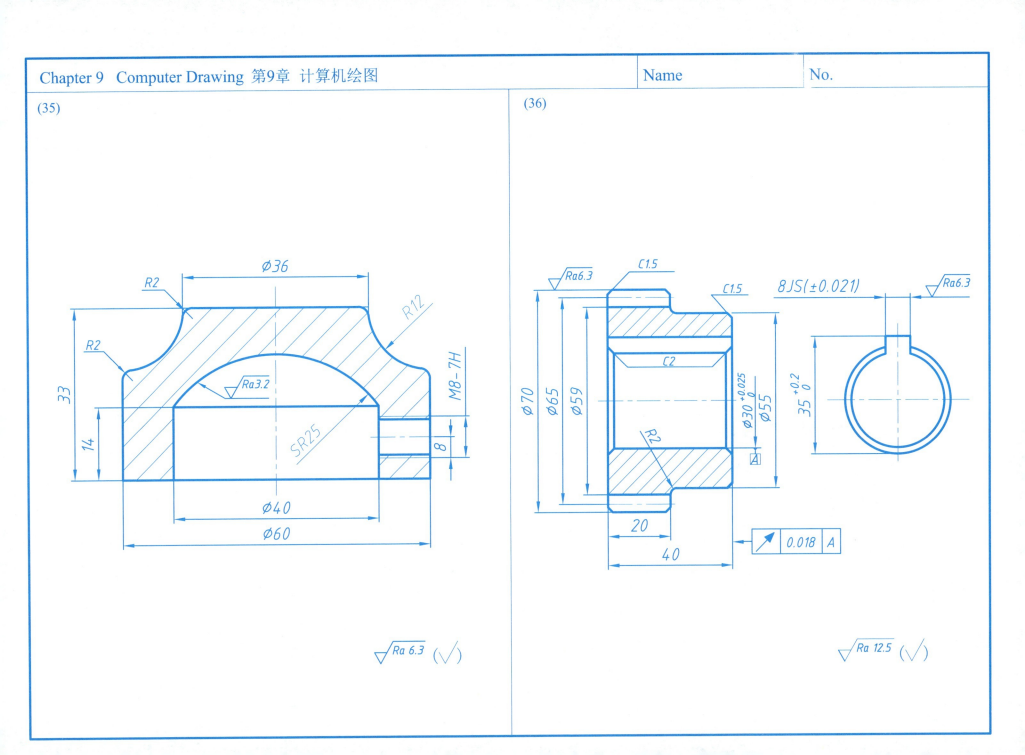

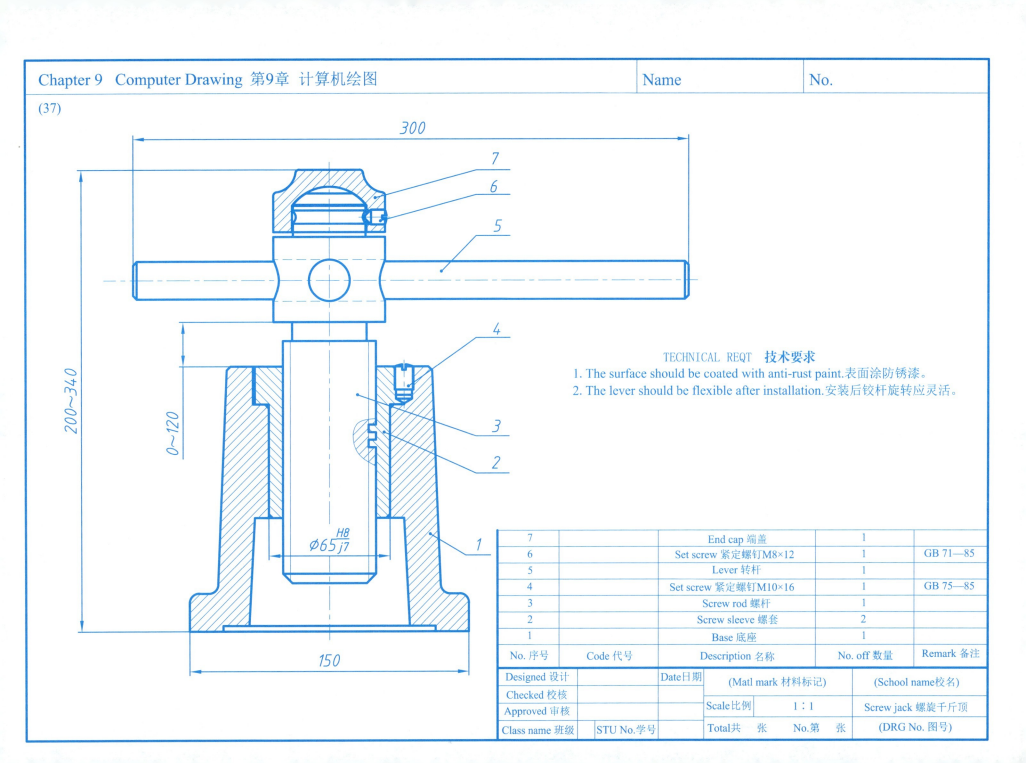

References 参考文献

[1] GRIFFITHS B. Engineering drawing for manufacture [M]. London: Kogan Page Science Ltd, 2003.
[2] SIMMONS C H, MAGUIRE D E. Manual of engineering drawing [M]. 2nd ed. London: Elsevier, 2004.
[3] 胡琳, 程蓉. 工程制图习题集（英汉双语对照）[M]. 3版. 北京：机械工业出版社, 2020.
[4] 大连理工大学工程图学教研室. 机械制图习题集[M]. 6版. 北京：高等教育出版社, 2013.